Handbook for Applied Modeling: Non-Gaussian and Correlated Data

Designed for the applied practitioner, this book is a compact, entry-level guide to modeling and analyzing non-Gaussian and correlated data. Many practitioners work with data that fail the assumptions of the common linear regression models, necessitating more advanced modeling techniques. This handbook presents clearly explained modeling options for such situations, along with extensive example data analyses. The book explains core models such as logistic regression, count regression, longitudinal regression, survival analysis, and structural equation modeling without relying on mathematical derivations. All data analyses are performed on real and publicly available data sets, which are revisited multiple times to show differing results using various modeling options. Common pitfalls, data issues, and interpretation of model results are also addressed. Programs in both R and SAS are made available for all results presented in the text so that readers can emulate and adapt analyses for their own data analysis needs.

JAMIE D. RIGGS is an adjunct lecturer in the Predictive Analytics program at Northwestern University, Chicago. She specializes in the statistical issues of solar system cratering processes, solar physics, and galactic dynamics, and has collaborated with researchers at the Los Alamos National Laboratory and the Southwest Research Institute. She has held technical and managerial positions at Sun Microsystems, Inc., National Oceanic and Atmospheric Administration, and the Boeing Company, where she applied advanced statistical designs and analyses to manufacturing and business problems. She is the head of the International Astrostatistics Association Solar System and Planetary Sciences Section.

TRENT L. LALONDE is Associate Professor of Applied Statistics at the University of Northern Colorado, and Director of the University's Research Consulting Lab. He has spent a number of years designing and teaching graduate courses covering statistical methods for students in diverse areas such as special education, psychological sciences, and public health. In addition, he has helped direct dissertations in these areas, and has consulted with numerous faculties on publications and funding proposals. He has received awards for both instruction and advising, and has chaired the Applied Public Health Statistics section of the American Public Health Association.

Handbook for Applied Modeling: Non-Gaussian and Correlated Data

Jamie D. Riggs
Northwestern University, Illinois

Trent L. Lalonde
University of Northern Colorado, Colorado

CAMBRIDGE
UNIVERSITY PRESS

CAMBRIDGE
UNIVERSITY PRESS

University Printing House, Cambridge CB2 8BS, United Kingdom

One Liberty Plaza, 20th Floor, New York, NY 10006, USA

477 Williamstown Road, Port Melbourne, VIC 3207, Australia

4843/24, 2nd Floor, Ansari Road, Daryaganj, Delhi – 110002, India

79 Anson Road, #06-04/06, Singapore 079906

Cambridge University Press is part of the University of Cambridge.

It furthers the University's mission by disseminating knowledge in the pursuit of education, learning and research at the highest international levels of excellence.

www.cambridge.org
Information on this title: www.cambridge.org/9781107146990

First published 2017

Printed in the United States of America by Sheridan Books, Inc.

A catalogue record for this publication is available from the British Library.

Library of Congress Cataloging-in-Publication Data
Names: Riggs, Jamie. | Lalonde, Trent.
Title: Handbook for applied modeling : non-Gaussian and correlated data /
Jamie Riggs, Northwestern University, Illinois, Trent Lalonde,
University of Northern Colorado.
Description: Cambridge : Cambridge University Press, 2017. |
Includes bibliographical references and index.
Identifiers: LCCN 2017004641 | ISBN 9781107146990 (hardback : alk. paper)
Subjects: LCSH: Mathematical statistics. | Mathematical models. |
Gaussian processes. | Stochastic processes.
Classification: LCC QA276.R5244 2017 | DDC 519.5/3–dc23
LC record available at https://lccn.loc.gov/2017004641

ISBN 978-1-107-14699-0 Hardback
ISBN 978-1-316-60105-1 Paperback

Additional resources for this publication at www.cambridge.org/riggslalonde

This book is dedicated to:

Lauren and Jordan. JDR
Amanda, for always listening. TLL

Contents

Preface   xiii

1   The Data Sets   1
1.1   Introduction   1
       1.1.1   The School Survey on Crime and Safety   2
       1.1.2   The Framingham Heart Study   2
       1.1.3   Fire-Climate Interactions in the American West   2
       1.1.4   English Wikipedia Clickstream Data   3
1.2   Exploratory Data Analysis   3
1.3   Gauss-Markov Assumptions   4
1.4   Data Summaries and Tables   4
1.5   Graphical Representations   4
       1.5.1   Histograms   5
       1.5.2   Q-Q Plots   5
       1.5.3   Box-Whisker Plots   5
       1.5.4   Scatter Plots   6
1.6   Pairwise Correlation   7
1.7   Machine Learning Pattern Recognition   7
1.8   Exploring the Data Sets   8
       1.8.1   School Survey on Crime and Safety Data   8
       1.8.2   Framingham Heart Study Data   13
       1.8.3   Fire-Climate Interactions in the American West Data   17
       1.8.4   English Wikipedia Clickstream Data   20
1.9   Summary   23
1.10   Further Reading   24

2   The Model-Building Process   25
2.1   Introduction   25
2.2   The Model-Building Process   26
       2.2.1   Exploratory Data Analysis   26
       2.2.2   Model Construction   27
       2.2.3   Model Fit Diagnostics   28
       2.2.4   Model Effects Analysis   28
       2.2.5   Model Interpretation and Prediction   29
       2.2.6   Effects and Predictive Model Differences   29
2.3   Constant Variance Response Models   30
2.4   Nonconstant Variance Response Models   31

2.5	Discrete, Categorical Response Models	32
2.6	Count Response Models	34
2.7	Time-to-Event Response Models	37
2.8	Longitudinal Response Models	39
2.9	Structural Equation Modeling	41
2.10	Effect Size	43
2.11	Model Fit Measures	43
	2.11.1 Measures of Fit	43
	2.11.2 Residual Analyses	45
2.12	Summary	48
2.13	Further Reading	49

3	**Constant Variance Response Models**	**50**
3.1	Introduction	50
3.2	School Survey on Crime and Safety	50
3.3	Framingham Heart Study	52
3.4	Fire-Climate Interactions in the American West	53
3.5	English Wikipedia Clickstream Data	55
3.6	Summary	56
3.7	Further Reading	56

4	**Nonconstant Variance Response Models**	**57**
4.1	Heterogeneity in Response Variance	57
4.2	Detecting Heteroscedasticity	58
	4.2.1 Descriptive Statistics	58
	4.2.2 Tests for Grouped Data	58
	4.2.3 Tests for Continuous Predictors	59
4.3	Variance-Stabilizing Transformations	59
	4.3.1 Selecting the Transformation	59
	4.3.2 Model Diagnostics	59
4.4	Weighted Least Squares	60
	4.4.1 WLS Estimation	60
	4.4.2 Selecting the Weights	60
4.5	SSOCS Analysis: Annual Suspensions	61
	4.5.1 Exploratory Data Analysis	61
	4.5.2 Normal Linear Model	63
	4.5.3 Outcome Transformations	63
	4.5.4 Weighted Least Squares	65
	4.5.5 Parameter Interpretations	68
	4.5.6 Model Prediction	69
4.6	Fire-Climate Analysis: Decade Averages	70
	4.6.1 Exploratory Data Analysis	70
	4.6.2 Normal Linear Model	71
	4.6.3 Weighted Least Squares	72
	4.6.4 Parameter Interpretations	74
	4.6.5 Model Prediction	74
4.7	Summary	75
4.8	Further Reading	75

5   Discrete, Categorical Response Models   76
5.1   Categorical Responses   76
5.2   Binary Logistic Regression   76
       5.2.1   Descriptive Statistics for Binary Outcomes   77
       5.2.2   The Logistic Regression Model   78
       5.2.3   Interpreting Model Coefficients   78
       5.2.4   Model Fit   79
5.3   Nominal Multinomial Models   81
5.4   Ordinal Multinomial Models   82
       5.4.1   Cumulative Logit Model   83
       5.4.2   Adjacent Categories Model   83
       5.4.3   Continuation Ratio Model   84
5.5   FHS Analysis: Probability of Hypertension   85
       5.5.1   Exploratory Data Analyses   85
       5.5.2   Logistic Regression Model   86
       5.5.3   Logistic Regression Model Fit   87
       5.5.4   Model Parameter Interpretations   89
       5.5.5   Model Prediction   90
5.6   SSOCS Analysis: Probability of Bullying   93
       5.6.1   Exploratory Data Analysis   93
       5.6.2   Ordinal Multinomial Model   94
       5.6.3   Ordinal Multinomial Model Fit   96
       5.6.4   Model Parameters Interpretations   97
       5.6.5   Model Prediction   99
5.7   Clickstream Analysis: Probability of Redlink   101
       5.7.1   Exploratory Data Analysis   102
       5.7.2   Logistic Regression Model   102
       5.7.3   Logistic Regression Model Fit   103
       5.7.4   Model Parameter Interpretations   104
       5.7.5   Model Prediction   105
5.8   Summary   106
5.9   Further Reading   107

6   Count Response Models   108
6.1   Introduction   108
6.2   Modeling Count Data   109
       6.2.1   Poisson Models   109
       6.2.2   Overdispersion   110
       6.2.3   Coefficient Interpretations   111
       6.2.4   Negative Binomial Models   113
       6.2.5   Zero-Inflated Models   114
       6.2.6   Zero-Deflated Models   114
       6.2.7   Hurdle Models   115
6.3   Fire-Climate Analysis: Decade Counts   115
       6.3.1   Exploratory Data Analysis   115
       6.3.2   Poisson Model   116
       6.3.3   Negative Binomial Models   118
       6.3.4   Zero-Inflated NB Models   119

6.4	SSOCS Analysis: Annual Suspensions	123
	6.4.1 Hurdle Negative Binomial Model	123
	6.4.2 Model Fit	124
	6.4.3 Model Interpretations	124
6.5	Clickstream Analysis: Site Pairings	126
	6.5.1 Exploratory Data Analysis	126
	6.5.2 Left-truncated Count Model	126
	6.5.3 Count Model Fit	128
	6.5.4 Coefficient Interpretations	129
6.6	Summary	130
6.7	Further Reading	131
7	**Time-to-Event Response Models**	132
7.1	Time-to-Event Data	132
7.2	Time-to-Event Models	133
7.3	FHS Analysis: Time to Hypertension	135
	7.3.1 Life Tables	135
	7.3.2 Kaplan-Meier Method	138
	7.3.3 Cox Proportional Hazards Models	140
	7.3.4 Time-Dependent Cox Models	145
7.4	Summary	150
7.5	Further Reading	150
8	**Longitudinal Response Models**	152
8.1	Longitudinal Data	152
8.2	Autocorrelation in Longitudinal Data	153
	8.2.1 Descriptive Analysis	153
	8.2.2 Scatter plots	153
	8.2.3 Autocorrelation Plots	154
	8.2.4 Variograms	155
	8.2.5 Modeling Longitudinal Data	156
8.3	Marginal Models	156
	8.3.1 Generalized Estimating Equations	157
	8.3.2 Working Correlation Structure	157
	8.3.3 Marginal Model Fit	159
8.4	Conditional Models	160
	8.4.1 Random-Intercept Models	160
	8.4.2 Random-Slopes Models	161
	8.4.3 Conditional Model Fit	162
8.5	FHS Analysis: Probability of Hypertension	163
	8.5.1 Exploratory Data Analysis	163
	8.5.2 Marginal Longitudinal Model	166
	8.5.3 Examining the Autocorrelation	166
	8.5.4 Marginal Longitudinal Model Fit	168
	8.5.5 Model Parameter Interpretations	168
	8.5.6 Model Prediction	170
8.6	Fire-Climate Analysis: Decade Counts	172
	8.6.1 Exploratory Data Analysis	172

		8.6.2	Autocorrelation in Decade Counts	175
		8.6.3	Conditional Models for Decade Counts	175
		8.6.4	Conditional Longitudinal Model Fit	176
		8.6.5	Model Parameter Interpretations	178
		8.6.6	Model Prediction	179
	8.7	Summary		181
	8.8	Further Reading		181

9 **Structural Equation Modeling** 183
9.1 Introduction 183
    9.1.1 SEM Variable Categories 184
    9.1.2 Model Types 185
    9.1.3 SEM Paths 185
    9.1.4 Confirmatory Factor Analysis 187
    9.1.5 Evaluating Model Fit 188
9.2 FHS Analysis: Latent Stress 189
9.3 SSOCS Analysis: School Climate and Academic Success 194
9.4 Summary 201
9.5 Further Reading 201

10 **Matching Data to Models** 202
10.1 The Decision Process of Modeling 202
10.2 Results of Model Application 207
    10.2.1 School Survey on Crime and Safety 207
    10.2.2 Framingham Heart Study 208
    10.2.3 Fire-Climate Interactions in the American West 208
    10.2.4 English Wikipedia Clickstream 209
10.3 Perspectives on Modeling 209

Bibliography 211
Index 213

Preface

Modern society is data driven. When you buy – or even shop for – a shirt on the Internet, the next time you enter the web, you'll be inundated with advertisements for more shirts, all the outcome of data collection, analysis, and targeted marketing. Global networks have been designed specifically to deliver stock market and commodities market data for near real-time trading. Public services depend heavily on censuses for allocation of government funding and assistance programs to the populations that need them. These same censuses determine the districts needed for so-called enfranchisement, at least in the United States. Travel, particularly international, is regulated based on personal information collected by government agencies. Large chain retailers collect cash-out data to stock according to collective shopping habits. Educators undertake quantitative assessments of new instructional methods to determine best practice. Health policy administrators analyze data to allocate resources according to the timing and volume of patient needs. These applications are just a hint of the universal use of data in both public and private spheres.

The ubiquity of data-driven decisions means that our personal and collective lives are affected daily by how data are analyzed and interpreted. When data are interpreted accurately, we expect fair treatment. When data are improperly collected, analyzed, or interpreted, not only is our quality of life diminished, but the faulty information can debilitate or even kill. Clearly, then, we want data analysts who, conscious of the consequences of poor or incorrect analyses, have the knowledge to extract information from data – properly and with a healthy awareness of any uncertainties that should qualify interpretation.

To support this kind of mastery, we have written this handbook to overcome two common limitations in tutorial resources for practicing data analysts.

- **We make a broad selection of the most useful basic models, from a range of disciplines and domains.** Applied disciplines that use statistical analysis sometimes rely on a restricted set of tools particular to the discipline. Although this practice has advantages at the entry level, it can encourage overreliance on familiar methods to the exclusion of viable, even superior, alternatives. This danger is compounded if discipline-specific software entrenches an unchanging set of models. Our approach is to look at a variety of data that is typical of modern applications and to present the models most likely to extract meaningful information. Our goal is not to present all possible useful models, but to build your facility with a range of core methods so that you are equipped to tackle new data with new or adapted models.

- **We deal with data as it comes, which is often non-Gaussian and often correlated.** Common practice, especially with large data sets, has been to assume that the data are close enough to Gaussian and uncorrelated even when these assumptions can be shown to be untrue. Misapplied analyses then produce tangles of misinformation. Our approach is to guide you to and through the statistical methods that best match the characteristics of the data under consideration, in particular methods suited to the prevalent non-Gaussian forms of observational data. Our goal is for you to become confident in building models for your real-world purposes.

This handbook is for data analysts with a grounding in basic statistics, biostatistics, econometrics, business statistics, social science statistics, or predictive analytics who want to develop their modeling skills beyond the commonly used, idealized setting of independent Gaussian analyses. We assume you are practiced in the use of descriptive statistics, analysis of variance, and regression.

All data analyses are performed on **real and publicly available data sets**, which are revisited multiple times to show differing results using various modeling options. You will see concrete examples of common pitfalls, issues that arise from messy data, and interpretation of model results. To encourage your hands-on engagement and so you can replicate any of the analyses, **code for all analyses is provided as both R and SAS** commands, available online at www.cambridge.org/riggslalonde.

The modeling methods are presented from a data analyst's perspective. We use basic mathematics to summarize model structure, basic model diagnostics, model effects interpretation, and predictive ability; however, our emphasis is always on the application of methods, rather than study of the methods themselves. We demonstrate how the methods are (or are not) appropriate, and how weak or strong is a model's performance with a given data set. This includes effects interpretation, predictive strength, and model aptness for model-to-model comparisons. While the book is well suited as a text for graduate-level methods courses, we present models through "standalone" discussions, so that you can use any single chapter as a self-contained resource for the models covered there.

Chapter 1 is the gateway to the rest of the book. In it the four driving data sets are described and explored, including all relevant variables used for analyses in later chapters. Chapter 2 gives a review of all the model types used in the book. Then, after a review of ordinary least squares estimation models in Chapter 3, we progress in the heart of the book through remedial methods under such violations of least squares assumptions as heteroscedasticity, serial correlation, and endogenous variables as found in panel data types, in addition to models for nonnormal responses, autocorrelated responses, time-to-event responses, and ending with structural equation modeling. The final chapter gives a point-by-point system for matching data to models.

Acknowledgements

We wish to thank Dr. Joseph M. Hilbe for sharing his vast knowledge and insightful wisdom. His encouragement in this endeavor was invaluable. Dr. Annyce Stone listened, commented, and encouraged us. Her top-down view gave further incentive to our work. Andrea Sorrells applied her artistic and critical eye to our cover art and images. We cannot thank her enough. Lucy Edwards of Cambridge University Press, New York, was patient, encouraging, cajoling, and, in general, a wonderful editor. She made this book a reality. Diana Gillooly believed in our data-centric approach to convey complex information.

1

The Data Sets

1.1 Introduction

This handbook is designed to provide an accessible introduction to statistical modeling techniques appropriate for data that are non-Gaussian (not normally distributed), do not have observations independent of each other, or may not be linearly related to selected predictors. The discussion relies heavily on data examples and includes thorough explorations of data sets, model construction and evaluation, detailed interpretations of model results, and model-based predictions. We intend to provide readers with a sufficiently thorough and understandable analysis process such that the techniques covered in this text can be readily applied to any similar data situation. However, it is important to understand that we use specific data sets with the various models strictly for demonstrative purposes. The outcomes we present are not to be assumed as definitive representations of information contained within the data sets.

Throughout the text, we will use four data sets (each described in this chapter) to exhibit the analytical methods including exploration of the data, building appropriate models (Chapter 2), evaluating the appropriateness of the models, output interpretation, and predictions made by the models. The purpose of using the same data sets throughout is to show that multiple methods can be applied to similar or identical variables of interest, possibly resulting in different conclusions. Consistent use of the same data sets should maintain data familiarity. After reading this first chapter, the intention of every data analysis throughout the remainder of the text should be understood. The modeling methods that are applied to the data sets are models for responses with constant variance (Chapter 3), responses with nonconstant variance (Chapter 4), discrete categorical responses (Chapter 5), models for count responses (Chapter 6), responses that are time-dependent (time-to-event data in Chapter 7, and outcomes collected over time in Chapter 8), and models for which variables that cannot be measured directly but are represented by variables that are measurable (Chapter 9). The last chapter, Chapter 10, is a guide to matching data sets to model types.

The following are brief introductions to each data set. These introductions are followed by descriptions of the data exploration methods that provide the details of the data sets needed to match them to the various models specifically designed for non-Gaussian and correlated data. Throughout the handbook, including the exploratory data analysis in this chapter, we use the functions and procedures, respectively, in the R (R Core Team, 2016) language and environment, and the SAS software, ©2016, SAS Institute Inc. SAS and all other SAS Institute Inc. product or service names are registered trademarks or trademarks

1

of SAS Institute Inc., Cary, NC, USA. All R and SAS commands used to produce the output discussed in this handbook are available.

1.1.1 *The School Survey on Crime and Safety*

The National Center for Education Statistics used the School Survey on Crime and Safety to record data from US public schools, addressing issues regarding safety in and around American public schools. Data for the 2007–2008 wave were collected from a stratified sample of 3 484 regular public schools. Variables of interest cover topics including school policies and facilities such as the presence of a school uniform policy and the use of metal detectors; school training and services such as teacher training on discipline policies and availability of counseling for students; and other variables such as the crime level of the areas surrounding school locations. We are interested in using these data to answer questions about school culture issues such as bullying and suspensions due to insubordination. The data can be downloaded from https://nces.ed.gov/surveys/ssocs/data_products.asp.

1.1.2 *The Framingham Heart Study*

The Framingham Heart Study was initiated to study common risk factors associated with cardiovascular disease, and to follow this disease's development over a long period of time for a large sample of participants. Data were collected from an initial cohort of 5 209 men and women between 30 and 62 years of age as of the time of a baseline physical examination. Data were also collected from two follow-up examinations performed two years and four years after the initial baseline measures. Variables of interest include patient demographic information such as age and sex; patient behaviors such as the number of cigarettes used per day; whether the patient uses blood pressure medication; and patient physiological measures such as systolic blood pressure, presence of diabetes, and body mass index. We shall analyze these data to answer questions about indicators of heart disease such as evidence of hypertension. The data can be downloaded from https://biolincc.nhlbi.nih.gov/teaching/.

1.1.3 *Fire-Climate Interactions in the American West*

Fire-Climate Interactions in the American West data since 1130 were obtained from the World Data Center for Paleoclimatology, Boulder, Colorado, and the National Oceanic and Atmospheric Administration (NOAA) Paleoclimatology Program (Trouet et al., 2010). These data were collected to assess whether climate is considered the main driver of wild fires in the American West. The data are on core samples from a variety of trees (Pipo, Psme, and Cade) from specific sites within each of the regions covering California, southern Oregon, and western Nevada. The regions are the Pacific Northwest (PSW), Northern California (NC), Interior West (IW), and the Southwest (SW). The time span is from 1130 through 2004. The core samples were examined to identify tree rings with fire scarring as an annual presence or absence of wild fires. We will use these data to predict indicators of fire scarring across regions and years. The data can be downloaded from www.ncdc.noaa.gov/paleo/study/10548.

1.1.4 English Wikipedia Clickstream Data

The clickstream data from the desktop version of the English Wikipedia were extracted from the logs of internet servers. The data are sequences of user-selected web addresses and links. Wikipedia (Wulczyn and Taraborelli, 2015) makes clickstream data available from its request logs, and we use the February, 2015 data. The data set includes only requests for articles in the main namespace. Pairings of the referring and requested sites with fewer than ten observations were removed from the data set by Wikipedia analysts.

The February, 2015 English Wikipedia Clickstream data set includes requests for redlinks (failed links), sorts out redirects, and has a field indicating whether the referrer and requested site pairings represent a link, a redlink, or a search. We intend to use these data to investigate the frequencies of pairings and the factors relating to redlinks. For examples of working with the February, 2015 release of the data, see this blog post: http://figshare.com/articles/ Wikipedia_Clickstream/1305770. The data can be downloaded from http://ewulczyn.github. io/Wikipedia_Clickstream_Getting_Started/.

1.2 Exploratory Data Analysis

Data are used to drive, support, or provide understanding of a wide variety of human activities. We often use existing data to suggest patterns that predict human behavior such as spending habits, health care access, voting outcomes, among many others. Available data are used to allocate expensive resources, or make decisions that may affect the quality of human lives, including survival. The costs of making erroneous inferences can be enormous. Understanding data collection methods, contents, and quality is a prerequisite to utilizing the data set for analysis. The first step in understanding the information contained within a data set is a thorough exploration of the set's variables' structures and contents. This exploratory data analysis (EDA) differs from data management, which is concerned more with data collection, organization, and quality attributes including accuracy, consistency, and completeness.

EDA, in the context of model-building, provides a guide as to what model types may be appropriate for the data set under consideration. A few common exploratory analyses are distribution investigation, frequencies of variable levels, variable correlations, and data summary statistics. Distribution investigation, when applied to prospective dependent or response variables, suggests whether they may satisfy the independence and distribution assumptions of specific models. Frequency analysis gives the number of levels within variables and by variables. These frequencies often suggest the use of, for example, indicator variables. Within- and across-variable correlations can indicate autocorrelation and possible issues with multi-collinearity. Data summaries offer measures of central tendency, ranges, and quantiles. These attributes help show if a chosen model's predictions are commensurate with the observed data.

EDA assists us in choosing models appropriate for a specific data set. EDA for model selection includes response-to-predictor relationships, predictor-to-predictor associations, response-by-predictor clustering, to name a few characterizations. A critical aspect of these characterizations is which distributions the data set variables may follow. The response variable distributions are particularly crucial to model type identification and selection.

The following is a review of common EDA methods and tools.

1.3 Gauss-Markov Assumptions

In statistics, the Gauss-Markov Theorem, named after Carl Friedrich Gauss and Andrey Markov, states that in a model with linearly related coefficients for which the errors have an expected value of zero, are uncorrelated, and have constant variance; the ordinary least squares (OLS) method produces the best linear unbiased estimators (BLUE) of the coefficients. Here "best" means the estimators have the smallest variance as compared to any other unbiased and linear estimators. The errors do not need to be normal, nor do they need to be independent and identically distributed. They need only be uncorrelated with mean zero and homoscedastic with finite variance. The requirement that the estimator be unbiased is mandatory as biased estimators may exist with smaller variance. Biased estimators can lead to unrealistic and unusable model outcomes.

In general we construct linear regression models under the expectation that the Gauss-Markov (G-M) assumptions hold, or that remedial measures may be taken to transform the data to approximate the G-M assumptions. EDA is a tool by which we may determine the compliance of such transformations to accommodate the G-M assumptions.

The following sections describe common EDA techniques that we shall reference throughout this handbook.

1.4 Data Summaries and Tables

Data summaries include descriptive statistics such as the mean, median, mode, range, minima, and maxima. These statistics generally are measures of central tendency and dispersion, variance, or spread. Continuous data may also be summarized using quartiles or percentiles. Partitions of this sort indicate how much grouping may exist in continuous data, and whether the two ends have a paucity or abundance of observations which indicate tail thickness. The values of the medians and means may suggest a variable has a symmetric distribution when the mean and median are equivalent, or skewed otherwise. Data scales are apparent from the minima and maxima.

Further summary of discrete and categorical variables may be made by generating two-way, three-way, or multi-way tables. For example, we may need to understand not just the counts of the levels within a categorical variable, but also the counts of various combinations of the levels of two or more categorical variables. Tables of categorical variable levels can contain not just the counts of levels, but also the percentages, fractions, or proportions based on the counts. As we shall see in Chapter 5, these tables may be used for assessing model fit.

1.5 Graphical Representations

The main graphical methods for exploring data are plots of distributions, response-to-predictor relationships, and predictor-to-predictor associations. The plots representing distributions include histograms, quantile-quantile (Q-Q) plots, and box-whisker plots. These plots suggest shape, including symmetry, modality, locations of central tendency, the amount of spread, and possible outliers. The most common plot for depicting associations between pairs of variables is the scatter plot. Machine learning such as ensemble learning utilizes clustering visualizations to depict variable grouping patterns. Ensemble learning and many other plot types for representing data behaviors are beyond the scope of this handbook.

1.5.1 Histograms

Histograms allow the examination of the distributional characteristics of a numeric variable using a specially constructed bar chart. Typically, a single variable is divided into groups, often called bins, the size of which defines the width of the bar. There are a number of ways these bins' widths can be determined, and we leave it to the reader to investigate the types of algorithms used by the software package being used. Once the bin width is set, the number or proportion of observations that fall within each bin is used to construct the height of the corresponding bar.

Each bar's height, as determined by the observation count, is divided by the total number of observations. The bar height now represents the fraction of the total number of observations centered on each bar within the width of the bar. How the adjacent bar heights are distributed will usually show symmetry or skewness, either to the left or to the right. The height of the left-most and right-most bars may suggest unusual tail thickness. A curve is often superimposed over the bars that represents a best fit probability density function.

The histograms suggest distribution characteristics, but when used in conjunction with descriptive statistics, other EDA plots, and distribution fit assessment statistics, they give information needed to choose which model will best fit the data.

1.5.2 Q-Q Plots

A Q-Q plot is a graph of one set of quantiles against another set. A quantile for any distribution; whether a normal, Poisson, or no apparent named distribution; is an element of equally-spaced ranks resulting from ordering the data from lowest to highest, followed by summing these ranks and dividing by the sample size. Often the data quantiles (the vertical axis, or ordinate) are plotted against the quantiles of a normal distribution (the horizontal axis, or abscissa). Deviations from, say, the normal quantile line suggest a non-Gaussian distribution.

1.5.3 Box-Whisker Plots

The box-whisker plot is a useful tool to partition a continuous or ordinal variable (e.g., a response for a model) into groups defined by some other discrete, prediction variables in the data set. The plot identifies, by group, asymmetries in the response, relative positions of the response quartiles, and possible extreme values. Each group's box-whisker plot is composed of five parts: (1) the box, (2) the horizontal median line inside the box, (3) a mean marker inside the box (though not a standard practice), (4) upper and lower whiskers extending from the box, and (5) indicators of observations above the upper whisker or below the lower whisker.

The five box-whisker plot parts are described as follows:

1. The vertical axis of the plot ranges from the smallest to largest values of the continuous or ordinal data variable. The horizontal axis ranks (generally, defined by the user) the order of the groups represented by the box-whisker plots. The box contains the portion of the range in which 50% of the data lie within a given group. The lower box boundary is the $Q_1 = 25$th percentile, and the upper boundary is the $Q_3 = 75$th percentile. The difference gives the range in which 50% of the data lie, and is known as the inter-quartile range (*IQR*). The width of the box sometimes is used as a conceptual measure of the sample size for each group.

2. Within each box is a horizontal line that represents the median value location of the data; viz., the 50th percentile location. Often this line connects notches with end points on the left and right sides of the box. The notches represent an approximate 95% confidence interval about the median value. The median value indicates that approximately half the data values lie at or below the median, and approximately half the data values lie at or above the median.

3. The mean marker is not always used, but when it is, it represents the location of the mean relative to the box. If the marker seems significantly shifted from the median line, an asymmetric distribution of the data is likely.

4. The upper and lower whiskers represent a bound in which approximately 90% of the data lie. The whisker lengths are found as $Q_1 - 1.5 \times IQR$ and $Q_3 + 1.5 \times IQR$, where *IQR* is as described in Part (1). Differing lengths of the upper and lower whiskers suggest nonsymmetric distributions.

5. Finally, the locations of values above and below the whiskers identify possible extreme values. Caution is advised before unequivocally designating these extreme values as outliers, as many probability distributions are skewed, and will allow their inclusion. The skewness results from the response possibly being from a nonsymmetric probability distribution. For example, count data often follow nonsymmetric distributions (e.g., Poisson), and the markers are not usually outliers.

1.5.4 Scatter Plots

Scatter plots represent the paired relationship between two variables. Three variables may be graphed in a single scatter plot, but they can be challenging to view with the possible exception of geographical plots. Geographical plots may show, e.g., a map of the United States with bars in each state representing a quantity such as health care costs. However, we focus on the relationships between pairs of variables as depicted in *x-y* plots; i.e., one variable is plotted on the horizontal axis and the other is plotted on the vertical axis.

Information depicted in two-way *x-y* scatter plots includes how one variable responds to changes in another, whether one variable has differing levels of variability at various levels or locations of another variable, and whether the levels of one variable tend to group on specific levels of another. Each observation is plotted as an *x*-axis and *y*-axis pair. The entirety of the observations plotted as *x-y* pairings gives a sense of the shape of the paired observations. The functional shape of the pairing may be depicted by a smoother such as the loess or spline smoothers. The shape is also dictated by the spread around a functional curve such as a straight line. The shape may be nonlinear, such as a quadratic function. The data points may appear equally spread around the function which suggests homogeneous variance.

When a variable is dependent on time, the variable-versus-time scatter plot is known as a time-series plot or a time plot. A time-series plot may show if the variable has constant variance through time, whether there is a trend such as a steady increase, or a nonlinear change through time such as a sinusoid. There is a class of statistical models designed specifically to analyze time-series data, but we do not consider it in this text.

Scatter plots, then, are a descriptive form of bivariate analysis. They suggest if a transformation of one variable results in a linear association rather than a curvilinear relationship, and whether a transformation converts nonhomogeneous variance to near homogeneous variance.

1.6 Pairwise Correlation

While the scatter plots give graphical representations between pairs of variables, the independence between these same pairs may be numerically evaluated using pairwise correlation. Pairwise Pearson correlation is a numeric measure of the linear association between two variables. Multiple variable pairs may be combined into a matrix for convenience. The correlation values are independent of scale (as opposed to covariance). This means that a variable with a large range can be correlated to a variable with a much smaller range, preserving the integrity of the correlation, even when the two ranges are measured with different units. It is critical to note that pairwise correlation has meaning only if the variable pair have a linear relationship.

Pairwise Pearson correlations range from $-1 \le \hat{\rho} \le 1$, where $\hat{\rho}$ is the estimated value of the correlation. (Often the correlation coefficient is denoted by r.) The closer $\hat{\rho}$ is to the extremes, the stronger is the linear correlation of a pair of variables. However large the absolute value of the correlation coefficient is, it may lack statistical significance due to such conditions as sample size. Therefore it is useful to also generate a significance level statistic such as a p-value.

Many data analysts suggest that only relatively large values of $\hat{\rho}$ should be used in, e.g., a linear regression model, when the correlation is between the model response variable and a candidate predictor variable. However, this reasoning is fallacious for three reasons: (1) the linearity of the correlation may be in question, (2) the pairwise behavior may change in the presence of variability due to multiple predictors, and (3) the intent behind an effects model and a prediction model are not always the same. Hence, we should always test for linearity in response-predictor relationships, we should never conclude a predictor has no influence on a response until it is tested in the multi-predictor environment if more than one predictor is used, and we must remember that effects models essentially identify the predictors that minimize model unexplained-outcome variation whereas predictive models may give more robust predictions if so-called pairwise noncorrelated response-with-predictors are included.

1.7 Machine Learning Pattern Recognition

Particularly for large data sets (the so-called big data), the methods of pattern recognition may prove beneficial. Pattern recognition methods are used in the discipline known as data mining, and include such techniques as cluster analysis, random forests, lift charts, regression trees, neural networks, nearest neighbors, and support vector machines, to name a few. Descriptions of these methods and techniques is beyond the scope of this book, but can be useful for identifying variable roles for modeling.

1.8 Exploring the Data Sets

We first examine the School Survey on Crime and Safety data, then explore the Framingham Heart Study data, followed by the Fire-Climate Interactions in the American West, and we finish with the English Wikipedia Clickstream data.

1.8.1 School Survey on Crime and Safety Data

We are interested in constructing models to make conclusions about general student behavioral problems, including bullying by students and suspensions for insubordination, using the 2007–2008 School Survey on Crime and Safety. Data were collected at the school level during the one-year period from 2007 to 2008, meaning that each variable represents an attribute of a school and not of any individual student. While the data set contains hundreds of variables, we have narrowed our focus to a few school characteristics.

- C0514: suspensions, the number of suspensions due to insubordination during the year.
- C0134: uniforms, an indicator of whether the school requires students to wear uniforms.
- C0116: metal detectors, an indicator of whether students must pass through metal detectors.
- C0188: tipline, an indicator of whether the school maintains a "hotline" or "tipline" for students to report problems.
- C0178: counseling, an indicator of availability of counseling or social work for students.
- C0562: crime, the crime level in the location of the school (low, moderate, or high).
- C0268: discipline training, an indicator of the availability of teacher training on discipline policies.
- C0276: behavioral training, an indicator of the availability of teacher training on positive behavioral interventions.
- C0508: insubordinates, the number of students involved in insubordination during the year.
- C0526: limited English, the percent of students with limited English language proficiency.
- C0532: below 15th, the percent of students who scored below the 15th percentile on standardized tests.
- C0376: bullying, how often student bullying occurs during the year in question (never, on occasion, monthly, weekly, or daily).

Continuous predictors of interest include the frequency of insubordinate students, the percentage of students with limited English language proficiency, and the percent of students below the 15th percentile on standardized tests. Noncontinuous predictors of interest include the level of crime in the area where the school is located (low, moderate, high), and indicators of whether students are required to wear uniforms, whether students pass through metal detectors, whether the school has a tipline to report problems, whether student counseling is available, whether teachers have training in discipline policies, and whether teachers have training in positive behavioral interventions.

Table 1.1 *School Survey on Crime and Safety descriptive statistics for continuous variables.*

Variable	Minimum	Median	Mean	Maximum	Variance
Number of suspensions	0	0	7.852	3 000	4 863.123
Number of insubordinate students	0	16	88.76	9 608	118 589.792
Percent with limited English	0.00	2.00	8.727	100	217.387
Percent below 15th percentile on tests	0.00	10.00	13.77	100	208.417

Table 1.1 shows basic descriptive statistics for the continuous variables of interest. Using this table we can see there are schools reporting 0 for each variable, and there are also schools reporting 100% for the two percentages. The number of suspensions and the number of insubordination events show clear evidence of skewness to the right, as the mean exceeds the median, and the maximum values, 3 000 and 9 608, respectively, are much greater than the median values. The median of 0 for the number of suspensions implies that at least half of the schools reported no suspensions during the year of interest.

The scatter plot matrix shown in Figure 1.1 gives a visual indication of possible relationships among the continuous variables. The histograms along the diagonal show evidence of skewness to the right in the percent of students with limited English language proficiency and also in the percent of students below the 15th percentile on standardized tests. Due to expected skewness to the right, both the number of suspensions and the number of insubordinate students were log-transformed (using the logarithm of suspensions+0.01 and the logarithm of insubordinates+0.01 to avoid undefined values from taking the logarithm of 0). The histograms from these two log-transformed variables show the smallest value to be the mode, which indicates that 0 suspensions and 0 insubordinates are the most common response for each.

The plot in the first row, second column of Figure 1.1 shows a reasonably strong relationship between the logarithm of suspensions and the logarithm of insubordinate students, as expected, and is supported by the relatively large value of 0.40 in the pairwise linear correlation. This scatter plot also shows an apparent diagonal "border" above which no observations are plotted. This is reflective of the fact that the number of suspensions does not exceed the number of insubordinate students. The remaining relationships appear relatively weak, and are affected by the large number of zeros for both number of suspensions and number of insubordinate students. In fact, the estimates of linear correlation between the percent of students with limited English language proficiency and both log-suspensions and log-insubordinates are so small as to not appear in the plot (0.006 and 0.005, respectively).

Table 1.2 shows the frequencies associated with each level of each categorical variable of interest in the data. Bullying has been recorded by schools as happening never, on occasion, monthly, weekly, and daily. The table shows most schools report bullying on occasion or monthly, but that daily bullying is more prevalent than no bullying at all. Most schools have no uniforms, no metal detectors for students to pass through, and no tipline to report issues,

Histograms, Scatter Plots, and Pairwise Correlations

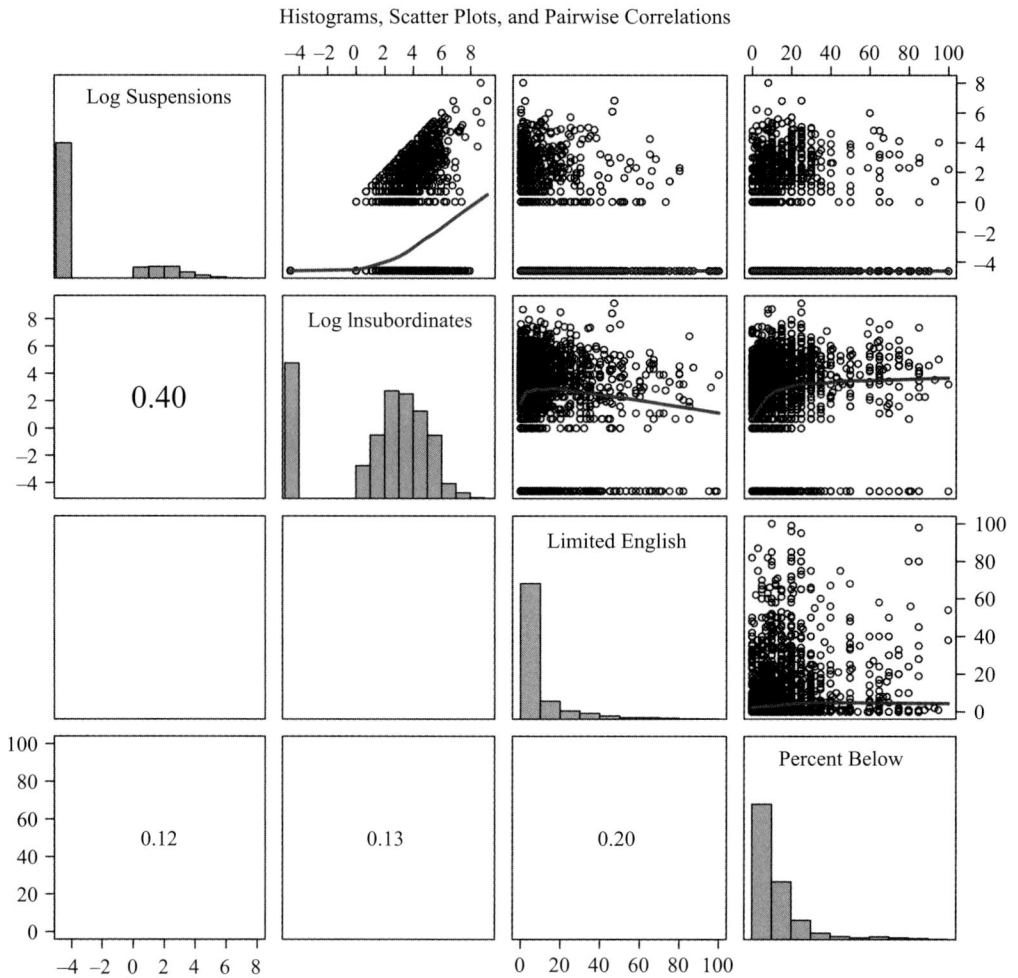

Figure 1.1 Scatter plot matrix of School Survey on Crime and Safety continuous variables, including histograms on the diagonal, pairwise Pearson correlations, and smooth loess curves.

but most schools have counseling available to students, are in locations of low crime, and have teachers trained in discipline policies and positive behavioral interventions.

In order to investigate the prevalence of predictor characteristics with school bullying, Table 1.3 shows cross-classification counts of schools that show combinations of bullying levels with the categorical predictors of interest. Counts can be used to describe patterns of variables of interest across bullying levels. For example, the "never" group has 30 schools with counseling and 10 without, a ratio of $30/10 = 3.00$, while the "on occasion" group increases to $1\,115/72 \approx 15.49$, and also ≈ 13.39, ≈ 21.64, and ≈ 23.00 for "monthly," "weekly," and "daily," respectively. The relative proportion of schools with counseling services available to students increases with frequency of student bullying; however, this change does not appear to follow a straight-line trend. Similarly, the proportion of schools in areas of moderate crime shows an increase across levels of bullying.

Table 1.2   *School Survey on Crime and Safety descriptive statistics for categorical variables.*

Variable	Levels	Number	Percent
Bullying	Never	40	1.6%
	On occasion	1 187	46.4%
	Monthly	547	21.4%
	Weekly	498	19.5%
	Daily	288	11.3%
Uniforms	Yes	377	14.7%
	No	2 183	85.3%
Metal detectors	Yes	65	2.5%
	No	2 495	97.5%
Tipline	Yes	901	35.2%
	No	1 659	64.8%
Counseling	Yes	2 406	94.0%
	No	154	6.0%
Discipline training	Yes	1 792	70.0%
	No	768	30.0%
Behavioral training	Yes	1 969	76.9%
	No	591	23.1%
Crime	Low	1 922	75.1%
	Moderate	494	19.3%
	High	144	5.6%

Table 1.3   *School Survey on Crime and Safety contingency table of categorical predictor counts, by bullying level.*

		Bullying				
		Never	On occasion	Monthly	Weekly	Daily
Total		40	1 187	547	498	288
Uniforms	Yes	10	174	83	66	44
	No	30	1 013	464	432	244
Metal detectors	Yes	1	35	10	12	7
	No	39	1 152	537	486	281
Tipline	Yes	12	405	190	184	110
	No	28	782	357	314	178
Counseling	Yes	30	1 115	509	476	276
	No	10	72	38	22	12
Discipline training	Yes	33	845	361	359	194
	No	7	342	186	139	194
Behavioral training	Yes	30	937	397	395	210
	No	10	250	150	103	78
Crime	Low	31	938	416	350	187
	Moderate	6	189	111	117	71
	High	3	60	20	31	30

Table 1.4 *School Survey on Crime and Safety descriptive statistics for number of suspensions, by categorical predictors.*

Variable	Levels	Suspensions average	Suspensions variance
Uniforms	Yes	8.777	1 407.280
	No	7.692	5 460.686
Metal detectors	Yes	19.185	4 006.028
	No	7.557	4 883.633
Tipline	Yes	12.646	12 011.913
	No	5.248	966.259
Counseling	Yes	8.079	5 150.157
	No	4.312	369.615
Discipline training	Yes	9.317	6 758.223
	No	4.434	427.555
Behavioral training	Yes	8.381	6 052.134
	No	6.090	901.265
Crime	Low	6.009	5 217.558
	Moderate	13.065	4 072.588
	High	14.569	2 708.778

Table 1.5 *School Survey on Crime and Safety descriptive statistics for continuous variables, by bullying level.*

		Insubordinates	Limited English	Below 15th
Bullying	Never	18.35(2 535.82)	10.48(357.13)	16.88(402.63)
	On occasion	64.95(33 029.27)	8.32(201.03)	12.77(201.62)
	Monthly	98.11(99 122.63)	8.46(226.71)	12.97(159.04)
	Weekly	104.20(138 104.02)	8.74(220.84)	15.10(215.89)
	Daily	152.24(485 167.10)	10.64(240.38)	16.69(275.67)

Table 1.4 shows descriptive statistics of the number of school suspensions for each group of the categorical predictors. Generally, the mean numbers of suspensions tend to be noticeably higher for schools with each of metal detectors, tiplines, counseling services, training on discipline policies and in areas of moderate or high crime. The differences in mean suspensions for uniform policies and positive behavioral training do not appear to be as large.

Table 1.5 shows means and variances of continuous predictors across levels of school bullying. Average values of insubordinate students increase as expected across levels of bullying frequency, although the percentage of students with limited English language proficiency remains relatively stable. The percentage of students below the 15th percentile in standardized testing shows some increase for weekly and daily bullying as compared to

Box Plots of Log-Suspensions by Crime

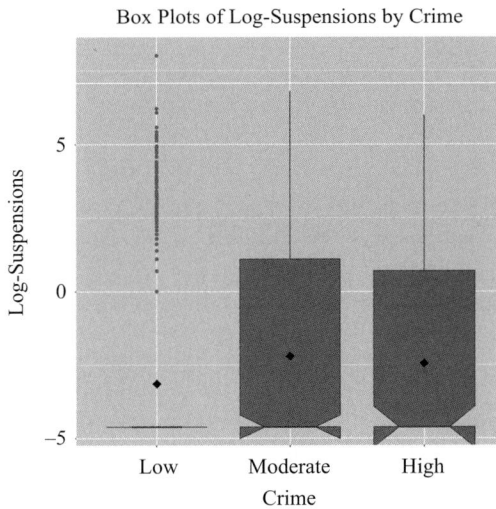

Figure 1.2 Box-whisker plots of log-suspensions for each crime level in the School Survey on Crime and Safety.

lesser levels of bullying. We can see that student bullying is not represented as a continuous variable, and therefore it would not be appropriate to model this outcome using normal linear regression techniques.

Figure 1.2 shows box-whisker plots of log-suspensions by area crime level. Schools in areas with low crime show the lowest typical numbers of suspensions, but also the largest number of possible outliers. The box-whisker plots of log-suspensions for schools in areas of moderate and high crime are similar to each other, but still show evidence of skewness to the right. Overall it appears the number of annual suspensions is highly skewed to the right and may have an excess of observations of 0, and therefore normal linear regression models would not be appropriate to model suspensions as an outcome.

1.8.2 Framingham Heart Study Data

We are interested in making conclusions about hypertension, including the prevalence of hypertension and the time until the onset of hypertension, using the data available from the Framingham Heart Study. The data comprise three waves of collection, each separated by two years. Some variables in the data set indicate measures specific to each wave, while others indicate presence or absence of a property over the entirety of the three-wave data set. For example, "PREVHYP" refers to each individual examination and may change over periods of observation, while "HYPERTEN" is an indicator of what was observed over the entire course of data collection. While there are numerous variables available in the public Framingham Heart Study data, we are concerned with only a few. Details of the processes involved with each physiological measure are available in the official documentation.

- RANDID: ID, a unique identification number for each participant.
- PERIOD: Period, examination cycle (1, 2, and 3).
- SEX: Sex, participant sex (1 for male, 2 for female)

Table 1.6   *Framingham Heart Study descriptive statistics for continuous variables.*

Variable	Minimum	Median	Mean	Maximum	Variance
Time of hypertension	0	2429	3599	8766	11996207.990
Total cholesterol	107.0	238.0	241.2	696.0	2059.309
Age	32.00	54.00	54.79	81.00	91.325
Cigarettes	0.00	0.00	8.25	90.00	148.000

- AGE: Age, the age, in years, of each individual at examination.
- TOTCHOL: Cholesterol, serum total cholesterol (mg/dL) at examination.
- DIABETES: Diabetes, an indicator of diabetic at examination.
- CIGPDAY: Cigarettes, the number of cigarettes smoked each day.
- PREVHYP: Hypertension, an indicator of prevalence of hypertension at examination.
- HYPERTEN: Hypertension, an indicator of whether each participant showed evidence of hypertension during any exam.
- TIMEHYP: Time to Hypertension, the number of days from the baseline examination to the first indications of hypertensive, or the number of days until the final contact is made with the participant if indications of hypertension are never recorded.

Continuous variables of interest include the time to hypertension, total cholesterol, and age. Noncontinuous variables of interest include sex, cigarettes per day, diabetic, and hypertensive. Table 9.2 provides basic descriptive statistics about each continuous variable. We note that some patients have a time of hypertension of 0 days, indicating presence of hypertension at the baseline exam. Both the minimum and the median of cigarette use are 0, indicating that at least half of the patients in the study do not use cigarettes on a daily basis, and suggesting this variable is skewed to the right.

The scatter plot matrix shown in Figure 1.3 gives a visual indication of possible relationships among the continuous variables. Many of the scatter plots fail to show evidence of strong relationships. Based on the plot in the first row, fourth column, it appears the most extreme values of total cholesterol are observed with the smallest times until hypertension. The plot in the first row, third column shows the variation in total cholesterol to decrease with the number of cigarettes smoked per day, and the plot in the second row, third column shows the number of cigarettes smoked per day to peak slightly before age fifty, and decrease with age. The small values of correlation in the lower panel of the figure support the lack of strong associations.

Histograms in Figure 1.3 show total cholesterol and age to be relatively symmetric, while cigarettes smoked per day is skewed to the right and time until hypertensive is bimodal, with common values toward the low end and high end of times.

Table 1.7 shows the proportions of specific outcomes associated with each categorical variable. The proportions suggest that around three-quarters of all exams resulted in evidence of hypertension, while only around 5% of exams showed evidence of diabetes. However, we cannot extend these simple descriptive statistics to make statements about percentages of patients, as the data were collected over time and represent multiple observations for each participant. Multiple observations on an individual, when ignored,

Histograms, Scatter Plots, and Pairwise Correlations

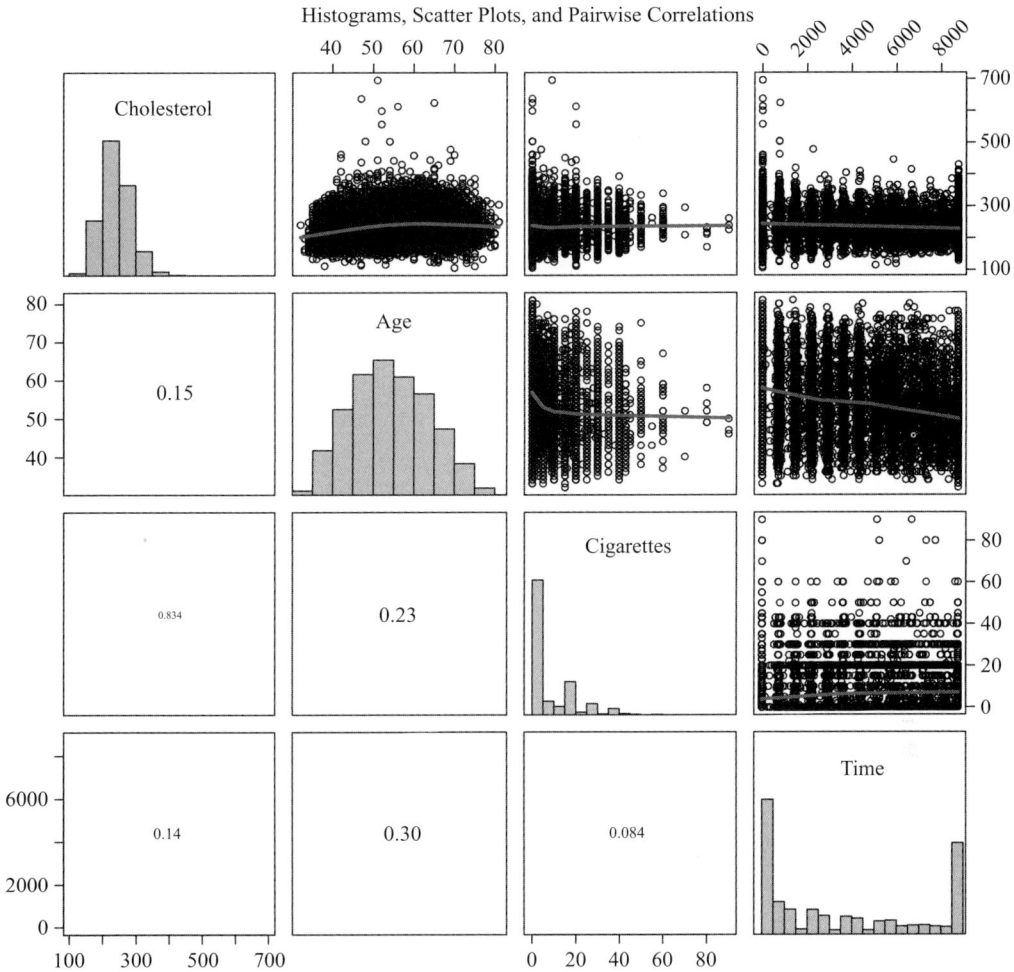

Figure 1.3 Scatter plot matrix of Framingham Heart Study continuous variables, including histograms on the diagonal, pairwise Pearson correlations, and smooth loess curves.

can bias the interpretations associated with descriptive statistics such as correlation and contingency table counts.

Table 1.8 shows cross-classified counts: the number of individuals with and without evidence of hypertension who showed diabetes and who were male or female. These splits in counts show us that there are more females than males with evidence of hypertension. Most of the individuals with diabetes show evidence of hypertension, although most of the individuals with hypertension are not diabetic.

Table 1.9 shows the descriptive statistics for the continuous variables of interest, split by hypertension. We see that individuals who show hypertension have much smaller values for the time of hypertension than individuals who do not show hypertension. This makes sense, as individuals who never show evidence of hypertension have time recorded as the final time of contact with the participant. Total cholesterol shows much greater

Table 1.7   *Framingham Heart Study*
descriptive statistics for categorical variables.

Variable	Levels	Number	Percent
Hypertension	Yes	8 642	74.3%
	No	2 985	25.7%
Sex	Female	6 605	56.8%
	Male	5 022	43.2%
Diabetes	Yes	530	4.6%
	No	11 097	95.4%

Table 1.8   *Framingham Heart Study contingency*
table of categorical predictor counts by hypertension.

		Hypertension	No hypertension
Total		8642	2985
Sex	Female	4956	1649
	Male	3686	1336
Diabetes	Yes	467	63
	No	8175	2922

Table 1.9   *Framingham Heart Study descriptive statistics continuous variables, by*
hypertension.

Variable	Minimum	Median	Mean	Maximum	Variance
	Hypertension				
Time of hypertension	0	826	2 116	8 764	6 360 355
Total cholesterol	107.0	241.0	243.8	696.0	2 087.197
Age	33.00	56.00	55.91	81.00	90.570
Cigarettes	0.00	0.00	7.674	90.00	146.881
	No hypertension				
Time of hypertension	45	8 766	7 891	8 766	3 543 448
Total cholesterol	117.0	229.0	233.4	430.0	1 894.903
Age	32.00	51.00	51.56	80.00	80.045
Cigarettes	0.00	3.00	9.924	80.00	149.561

variation for participants who show hypertension, as evidenced by the smaller minimum and larger maximum than individuals without hypertension. Individuals without hypertension generally show greater cigarette use, as evidenced by the larger mean and median than those with hypertension.

Figure 1.4 shows plots of hypertension ("1" represents "yes" and "0" represents "no") versus age and total cholesterol. Because the outcome of interest, hypertension, is binary, it is difficult to identify a pattern as with a typical scatter plot. Therefore the loess smoothed curve is superimposed to show the general increasing relationship between each predictor

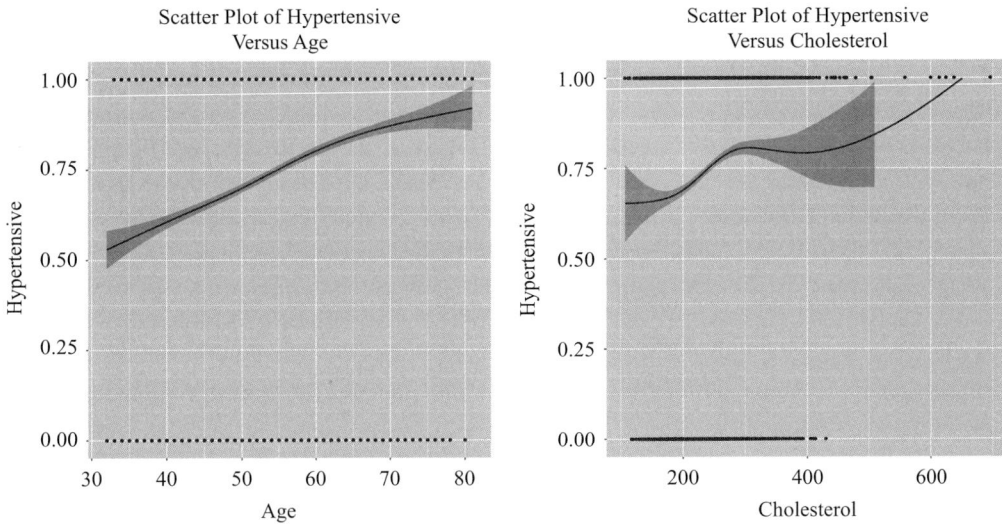

Figure 1.4 Framingham Heart Study plots of hypertension versus age (left panel) and total cholesterol (right panel), with smooth loess curves.

and the prevalence of hypertension. In this case the loess curve can be thought of as a smoothed estimate of the proportion of individuals showing evidence of hypertension for various values of the predictors. It appears the proportion of individuals with hypertension increases with both age and cholesterol, although the nature of the relationship does not appear to be linear for cholesterol. The exploratory statistics presented suggest the relationship between hypertension and predictors such as age and cholesterol should not be investigated using normal linear regression methods.

1.8.3 Fire-Climate Interactions in the American West Data

The models of Fire-Climate Interactions in the American West use the number of annual tree ring fire-scar markers for the years beginning 1130 through 2004. The regions and the sites in which the tress are located are used to construct the models. The tree ring fire-scar markers from individual trees that experienced fires since the year 1130 are used. These marker tree rings were collected from among the 350 sites in each of the four regions of the fire data, but we are concerned here only with the sites listed below (Trouet et al., 2010).

- year: the year indicated by sampled tree rings.
- region: four regions are used in this text. They are the Intermountain West (IW), Northern California (NC), the Pacific Northwest (PNW), and the Southwest (SW).
- site: within region are sample sites. They are:

  - IW: Ashenfelder, Cheesman Lake, Manitou, and Old Tree
  - NC: sites not used
  - PNW: Frosty, Nile Creek, South Deep, and Twenty Mile
  - SW: Blacks, Round Mountain, and Cerro / Hoya / Marchanita

Table 1.10 *Fire-Climate continuous variables data summary.*

Variable	Minimum	Median	Mean	Maximum	Variance
Decade	1130	1604	1610	2000	54825.246
Year	1130	1608	1604	2004	54810.000
Rings/year	0	0	0.756	19	3.061
Rings/decade	0	28	30.55	87	614.120

Table 1.11 *Fire data site by region contingency table (the NC sites are not used).*

	IW	NC	PNW	SW	Total
Total	875	841	661	771	3148
Ashenfelder	88		0	0	
Blacks	0		0	78	
Cerro Hoya Marchanita	0		0	78	
Cheesman Lake	88		0	0	
Frosty	0		67	0	
Manitou	88		0	0	
NileCreek	0		67	0	
OldTree	88		0	0	
Round Mountain	0		0	78	
Twenty Mile	0		67	0	

The fire-scarred tree ring samples were taken from the American West, stratified by region and further stratified by site. Models may pool over site or over region, or models may account for the stratifications through longitudinal analysis. This handbook accrues the yearly counts into decadal summaries such as means or counts prior to applying these data to various model types.

The fire data continuous variable summary is given in Table 1.10. The row "decade" summarizes the number of decades from between 1130 and 2004. The column labeled "region" gives the numbers of tree ring samples in each of the four given regions. "Year" is the number of years between 1130 and 2004. The counts of tree ring samples by year are given in "rings/year," and the number of tree ring samples aggregated by decade is in column "rings/decade." Table 1.11 has the counts of rings for site by region. The Northern California region's sites are not used, and hence the associated site counts for region NC are not given.

Figure 1.5 is a matrix plot of the decade, decadal counts, and the log-transformed decadal counts which allows us to examine these variables' histograms, pairwise scatter plots (with loess smoothers), and the pairwise linear correlations. The plot matrix diagonal gives each variable's histogram. The first row, first column histogram is of decade, which is of minor interest.

The second row, second column panel is the decadal counts in which we see that smaller counts dominate. Smaller counts also dominate in the third row, third column panel of the matrix which is the histogram of the log of the decadal counts. Note that for purposes of

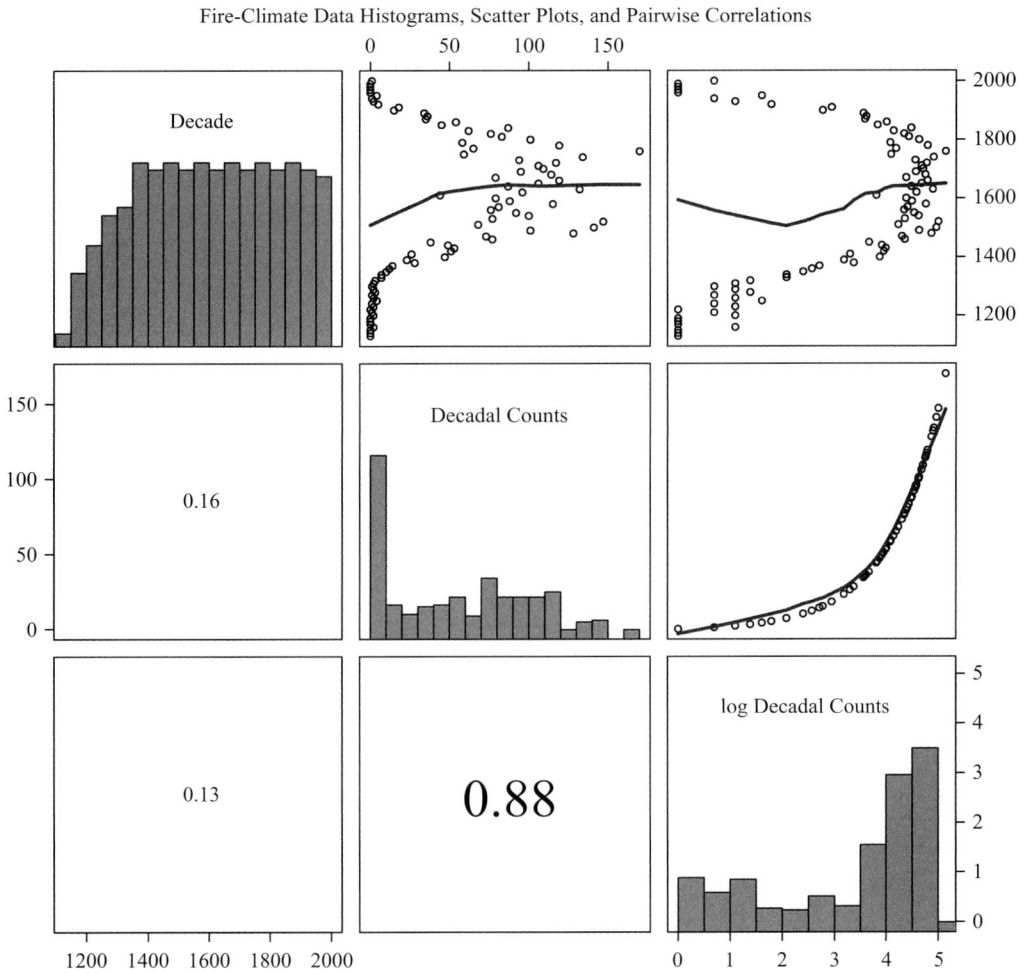

Figure 1.5 Matrix plot of the histograms, scatter plots, and linear correlation statistics of the three pairings of decade, decadal counts, and log-transformed decadal counts.

generating the log plot a value of one is added to each count as the raw counts include the value zero. This histogram shows a shift of dominance from the low counts of the untransformed decadal counts, to the higher values of the transformed data. Even though the transformation has shifted the mode of the distribution, it is clearly not a Gaussian distribution.

The linear correlation value for the pairing of decade and the decadal counts is the second row, second column panel of Figure 1.5. The small correlation, $r = 0.16$, suggests no linear relationship. The linear correlation value of decade and the log of the decadal counts is given in the third row, first column. The small correlation, $r = 0.13$, suggests no linear relationship.

The upper triangular matrix of cells of Figure 1.5 are scatter plots of the pairings. We are not interested in the pairing of decadal counts and log of decadal counts except, if there is interest in the appearance of the log transformation. We see that neither decadal count (first row, second column) nor log decadal count (first row, third column) is a linear function

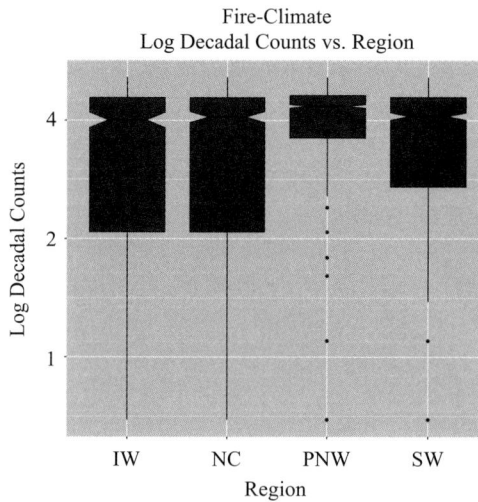

Fire-Climate
Log Decadal Counts vs. Region

Figure 1.6 Box-whisker plot of the log of decadal count by region.

of decade. This nonlinear behavior invalidates the correlation statistics, as does the lack of normality in the histograms, and we must make an accounting of these nonlinearities when constructing models.

The fire data summary, Table 1.10, shows that region is a categorical variable with four levels, and for each level, the table gives the number of tree ring samples. The table shows that no region has zero counts, allowing each region to be included in models. We examine the log decadal counts by region using a box-whisker plot in Figure 1.6. Note the skewness in each of the plots, which is a reflection of the combined log counts skewed histogram we saw above. This suggests a Gaussian model is likely an inappropriate model type for these data.

1.8.4 English Wikipedia Clickstream Data

Wikipedia (Wulczyn and Taraborelli, 2015) makes clickstream data available from its request logs, and we use the February, 2015 English subset of these data. The link pairings of website referrals and site requests are examined by link type and previous site. The data are requests for articles in the main namespace of the desktop version of English Wikipedia. Pairings of the referring and requested sites with fewer than 10 observations were removed from the data set by Wikipedia analysts.

The data consist of six variables and we use the following three, whose definitions are from Wulczyn and Taraborelli (2015):

- n: the number of pairings of the referring and requested sites indicating the number of times web paths are used.
- prev_title: the mapping of a referring URL to any of the following values: Bing, empty, Google, Main_Page, other, Wikipedia, or Yahoo.
- type: indicates if the pairings of the referring and requested sites are:

Table 1.12 *English Wikipedia Clickstream continuous and categorical variables descriptive statistics. The number of pairings is the continuous (counts) variable. Link type and previous title are the categorical variables.*

Variable	Minimum	Median	Mean	Maximum	Variance
Number of pairings	10	95.05	23	95 738	778 093.6

Variable	Levels	Number	Percent
Link type	Link	37 545	57.35%
	Other	27 602	42.16%
	Redlink	317	0.48%
Previous title	Bing	1 593	2.43%
	Empty	6 733	10.29%
	Google	8 078	12.34%
	Main_Page	509	0.78%
	Other	43 359	66.23%
	Wikipedia	3 962	6.05%
	Yahoo	1 231	1.88%

- link: the referrer and request are both articles and the referrer links to the request.
- redlink: the referrer is an article and links to the request, but the request is not in the production enwiki.page table.
- other: the referrer and request are both articles but the referrer does not link to the request.

A summary of the clickstream data is given in Table 1.12. The summary includes of the number of pairings of the referring and requested sites. The table gives the counts of the three levels of link type ("link type") and the counts of the seven previous titles' ("previous titles") levels. The table shows that no category has zero counts, allowing each category to be included in models.

The combination of the pairings of link types and previous titles is given in Table 1.13. This two-way table is important to determine if any combination of the category levels has zero or small numbers relative to the expected cell count. Small counts tend to dominate from expected cell counts, thus possibly biasing tests of significance. The table shows zeros and counts of 1 and 2 for several combinations of the levels, and hence, interacting these two categories as a predictor in a model is likely to fail.

The numbers of pairings will be partitioned in Chapter 5 for modeling as the probability of these pairings. In Chapter 6, the counts of pairings will be modeled.

Figure 1.7 shows the histograms of the counts of referring and requested pairings. The left-hand panel is the raw counts, and we see that smaller counts dominate even though it was necessary to exclude all counts above 1 000 to obtain a displayable plot. The right-hand panel of the figure is a histogram of the natural log of these same counts, and it is clear that the data are truncated on the left. This is verified by the minimum count in Table 1.12 at 10. Each plot has an overlay of a smoothed density curve.

The relationship between the categories in Table 1.12 and the log counts is shown in Figure 1.8. The left-hand panel of the figure is a box-whisker plot of the pairings types

Table 1.13 *English Wikipedia Clickstream link type by previous title contingency table.*

Previous title		Link type			
		Link	Other	Redlink	Total
	Total	37 545	27 602	317	65 464
Bing	Count	1	1 592	0	1 593
	Percent	30%	1 013%	0.00%	
Empty	Count	2	6 731	0	6 733
	Percent	0.00%	10.28%	0.00%	
Google	Count	2	8 075	1	8 078
	Percent	0.00%	12.34%	$< 0.01\%$	
Main_Page	Count	2	507	0	509
	Percent	0.00%	0.77%	0.00%	
Other	Count	37 537	5 506	316	43 359
	Percent	57.34%	8.41%	0.48%	
Wikipedia	Count	1	3 960	0	3 961
	Percent	0.00%	6.05%	0.00%	
Yahoo	Count	0	1 231	0	1 231
	Percent	0.00%	1.88%	0.00%	

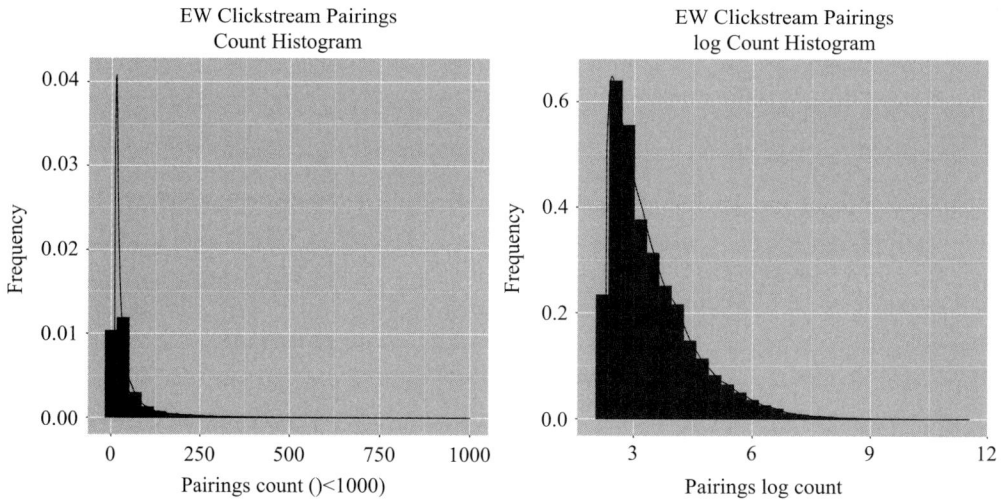

Figure 1.7 A histogram of the English Wikipedia Clickstream raw counts of the referrer and requested sites pairings counts is given in the left panel. Note that in order to produce a viewable plot, counts greater than 1 000 were excluded. The right-hand panel is a histogram of the log of the pairings counts. Both plots have smoothed density overlays.

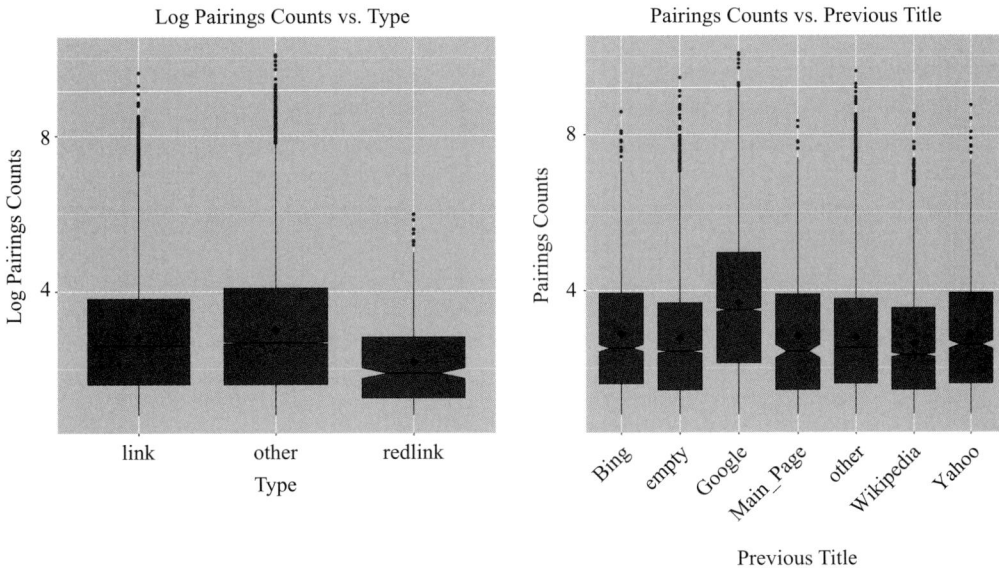

Figure 1.8 English Wikipedia Clickstream link type log counts (left panel) and the log counts of the previous titles (right panel). All the categories show skewed behavior.

(Type) log counts, and the right-hand plot is a box-whisker plot of the pairings of Previous Titles log counts. Note the skew in each of these plots. This suggests a Gaussian model is likely an inappropriate model type.

The EDA for the clickstream data show two important points to consider when choosing and constructing a model. The first is that the number of pairings fails to follow a Gaussian distribution function, and hence, normal linear regression is likely to be an inadequate and inappropriate model. The second point is that the categorical variables type and previous title show several zero counts in the two-way table which means either the interaction of the two variables will fail as a predictor, or remedial measures such as combining levels must be made if an interaction is to be used. Subject-matter expertise should be used prior to implementing remedial measures to assure conformance to the study intention.

1.9 Summary

Data sets usually contain variables with missing values and possible outliers. Rarely are they as well behaved as those in many textbooks, specifically, textbooks on empirical modeling methods. Messy data are seen in some of the tables and plots in this handbook. EDA is not only critical for finding and constructing optimal models based on response distributions, it is mandatory for identifying unbalanced data, missing values, sparse cross-classifications of variables, and the like. Possibly severe consequences can result if EDA is ignored as a precursor to model identification and construction. Throughout this handbook we will make reference to the exploratory analyses presented in this chapter, and we will build upon these basic descriptions as we construct models for specific data situations.

1.10 Further Reading

John Tukey elucidated the essence of EDA in Tukey (1977). The EDA techniques we have presented in this chapter are but a subset of the methods presented by Tukey's text.

Cleveland (1993) demonstrates the visualization of data including user interaction and animation. This book focuses on interfaces around large and complex data sets.

Good introductory texts on basic statistical methods include McKillup (2011), Freedman et al. (2007) and Agresti and Finlay (2008).

2

The Model-Building Process

2.1 Introduction

Model types vary widely depending on the type of data under consideration and the purpose of an investigation or analysis. In this chapter we give a short introduction to the statistical models used in this handbook. The reader may go directly to the chapter on how to use the model of interest. However, as the reader becomes more adept in model application, this chapter will be useful for acquiring additional model knowledge, and for finding literature to enhance the flexibility of and appreciation for the various models discussed.

Models, either empirical or mechanistic, allow us to gain insight into usually complicated processes. Models are used to represent particular aspects of a hypothesis, study subject, or process. As examples, a toy airplane represents the overall concept of a type of flying machine as it has the general shape and appearance of an aircraft. A hobby plane may represent intricate appearance details of a specific plane, but does not fly. A flying model may have the general appearance of a particular aircraft, but it generally lacks details of actual airplane appearance and instead is optimized specifically to fly. Then there are simulators which do not fly but allow for a facsimile of flight with performance characteristics intended to mimic the actual flying machine to train pilots. We see from these examples that each model has a different purpose, each with different relative advantages and disadvantages.

Models, then, have deficiencies relative to the object or process they represent. However, models do provide insight into a particular aspect of these objects or processes, and thereby they are useful. Models of gravitational attraction are mechanistic as nature dictates the functional relationship between, say, two masses. Stochastic or empirical models are often used to discover functional relationships among variables. They are used to provide summary information about sets of data. While it is possible to examine each observation in a data set, as the size of the data set increases, it is evermore difficult to assimilate the information held by each datum. A model matched to the data set can reveal processes and functional relationships otherwise inaccessible to individual observational examinations.

Empirical models are often thought to describe causal relationships between a response variable and a set of predictor variables. However, only very specifically designed studies may describe causality. Generally, we consider most empirical models to describe only the influence, association or relationship between the predictors and the response, both of which are observed. The predictor and response variables can be said to be related, and associated covariation can be expected, but the additional step of declaring one variable to *cause* a change in the response is typically not supported by observational designs. This may seem a subtle difference, but is a critical one. The data sets used in this handbook are observational, and hence causal relationships between responses and predictors are proscribed. The empirical model construction must associate the observed response with the observed predictors by using these observations to estimate coefficients that make this association. It is important to note that *predictions* can be made using observational data, that is, model prediction does not necessitate a causal relationship. Predictions can be made based on observed relationships without declaring the cause of each relationship.

The model-building process is usually iterative in that once a model is constructed, it may be necessary to rebuild or adjust the model to account for information discovered during model evaluation. This implies that we must decide what model type is suitable, and we must decide if the constructed model is adequate for the intended purpose. Specific sets of data called pilot data often are employed to reduce the numbers of possible model choices. In fact, pilot data are an often overlooked necessity to performing the circular process of updating models based on characteristics of a given data set. As an alternative, data analysts often perform a process of selecting a small percentage of a single data set to be used for model selection, as a pilot data set. The selected model is then applied to the remaining, unused portion of the original data set to answer scientific questions of interest. In this way analysts avoid the questionable process of using a single collection of observations to both select an appropriate model and to answer questions of interest using that model, which often represents the single data set well but does not effectively generalize to other situations. Most importantly, the model must provide information that permits insight into, and interpretation of, the nature of the process or study goals as revealed from the data.

2.2 The Model-Building Process

We consider that the model-building process consists of five components, which we will use to construct the various models presented in this handbook. The components are:

1. Exploratory data analysis (EDA)
2. Model construction
3. Model fit diagnostics
4. Model effects analysis
5. Model interpretation and prediction

Each component is explained in the following subsections.

2.2.1 Exploratory Data Analysis

Exploratory data analysis, in the context of model building, provides a guide to what model types may be appropriate for the data set under consideration. A few common exploratory analyses, described in Chapter 1, are distribution investigation, frequencies of variable levels, variable correlations, and data summaries. Distribution investigation, when applied to the dependent or response variables, suggests whether they may satisfy the normal regression model assumptions including independent, Gaussian responses with constant variability. Frequency analysis gives the number of levels within variables, and by variables. These frequencies often suggest the use of, for example, indicator (dummy) variables. Response and predictor combinations within and across correlations can indicate autocorrelation and possible issues with multicollinearity. Data summaries offer measures of central tendency, ranges, and symmetry.

EDA is critical to choosing a model that matches to the data and that is commensurate with the analytical objectives. A model response variable that is non-Gaussian very often results in normal linear model residuals that violate the other Gauss-Markov assumptions, including a linear relationship and constant variation. Predictor combinations may be unstable for the various parameter estimation methods such as ordinary least squares or maximum likelihood. EDA can guide us to choose an appropriate model, as well as give clues to what issues the fitted model must address.

2.2.2 Model Construction

The EDA and analysis objectives suggest the model type or types necessary to satisfy the study objectives and indicate a match to the data. It is often necessary to construct more than one model, as the study objectives may require multiple models, or multiple model versions that are improvements from one model to another. Model construction, then, is usually iterative, and therefore comparative measures of model adequacy must be assessed and interpreted. Best practice in reporting on final model adequacy calls for comments on why multiple models are used, and in the case of model versions, levels of improvement are useful.

Model specification is required before a model is constructed. Specification involves listing the model assumptions, the response-to-predictor relationships, and predictor configurations. The assumptions include the expected unexplained variance (through residuals) behavior and the mean and variance relationship of the response as, say, a statement about the distribution of the response. The main model structural concern is the functional relationship between the response and the predictors. The functional relationship is known as a link function, with common options being the identity function and the natural exponential function. The predictor configuration involves adjusting for possible across-predictor correlations (centering is a remedy), transformations to account for nonlinearity between a predictor and the response, and identifying possible outliers for remediation and explanation. If the specifications are not understood or if they are neglected, the choice of model may be compromised. In such a case, outcomes generated by the model are likely erroneous, and thus interpretations are at best questionable.

Once a model is specified, a method for estimating the model parameters must be chosen. The model specification generally mandates the estimation method, but in many cases multiple estimation options are available. For example, if the EDA suggests a normal linear model is an appropriate match to the data and study objectives, the most efficient estimation method is ordinary least squares. If the EDA and model specification includes random effects predictors, then a form of maximum likelihood estimation may be optimal. Iterated (re-)weighted least squares (IWLS or IRLS) is a common computational algorithm to calculate maximum likelihood estimates for discrete and count response models. Most statistical modeling packages allow the user to choose from estimation method options.

Using iterative estimation methods such as IRLS or other nonlinear approximations to maximum likelihood, it is possible that the characteristics of the data will cause the estimates produced by this process to fluctuate and fail to produce single, final estimated values for parameters. In such cases we say the estimation method failed to converge, and an alternative method of estimation must be pursued or the model must be adjusted. When a model has no estimation problems such as inversion or convergence failures, model diagnostics and fit measures must be examined to assure the model provides usable outcomes.

2.2.3 Model Fit Diagnostics

The sufficiency and conformance of a model to the model specification, model fit to the data, and adherence to the analysis objectives are assessed in part with model diagnostics. These diagnostics take the form of graphical representations of, particularly, remaining unexplained variation as well as measures of fit. The graphical representations can identify nonhomogeneity of variance, residuals distributions, and conformance of the model-fitted data to the observed data. The fit measures not only show the efficacy of the model fit, but also allow for model version comparisons. Diagnostics are mandatory on all models that are referenced for final interpretations.

We give more detail on model fit measures in Section 2.11. We discuss the various diagnostic plots and the variety of fit statistics which are often dependent on the model type and specification.

2.2.4 Model Effects Analysis

Once a model is constructed, the estimated model parameters must be evaluated for significance, level of contribution to variance reduction, and the magnitude of the effect on the response. Without these assessments, model evaluation is incomplete. Usually the study objectives will mandate an evaluation of the effects on the response. A large effect size of one variable can dominate response changes even though other effects are significant.

Parameter significance is a function of the value of the parameter estimate and the size of the parameter's standard error. Depending on the model type, the significance statistic may be based on a normal z-statistic, a t-statistic, or a chi-squared statistic. Associated with these statistics is a value of the probability of the statistic having a more extreme value and hence more significant level. This probability value is known as a p-value, and represents an estimate of the likelihood of the observation of certain data characteristics in future studies. The study must include an *a priori* statement of the level at which a p-value is deemed small enough to represent a significant effect, but note that p-values do not measure the probability that the studied null hypothesis is true, nor the probability that the data were produced by random chance alone. Scientific conclusions and business or policy decisions should not be based solely on whether a p-value passes a specific threshold as p-values do not uniquely determine meaningful associations nor substitute for subject matter expertise.

The effect a predictor has on a response is not measured by the associated p-value alone, rather, a level change in the predictor is multiplied by its estimated coefficient to register the change in the response. If multiple predictors are used in a model, the values of all but the predictor of interest are held at specified values such as predictor means. If the predictor of interest is assumed to change by unity, then the coefficient directly represents the response change. If more than one version of a model is constructed, effects analyses are not always needed until a final model version is determined, but effects analysis is necessary on any final models. Effects analysis allows us to quantify and depict the change in a response by any one predictor, or, if predictors interact, then the response change is due to one predictor's change at set values of the interacting predictor.

Effects analysis informs us as to how an expected response changes relative to changes in the predictors. If the changes are contrary to common-sense changes of the subject matter, more investigation into the selected model and even to the data is warranted.

2.2.5 Model Interpretation and Prediction

Constructing models to answer research questions or satisfy study objectives are of little use if the outcomes are not given interpretations and conclusions are not drawn. Model outputs alone are the basis of interpretation, but are far from complete. All analysis outcomes require deriving interpretations and conclusions relative to the subject matter of the study. Some elements of the interpretation include confidence in the model, confidence in the model predictions, and model performance relative to study objectives.

Confidence in the model is derived from the model fit statistics and model diagnostics. Confidence in the model predictions is determined by use of cross-validation of training data sets and test data sets, or comparison with existing, similar models. Model performance relative to study objectives is accomplished by assessing predictions against realized, future observations.

Model outcome interpretation is the reason for construction of the model. The model's purpose is to reveal often hitherto unknown relationships among a set of variables as represented by a data set. The model must allow us to add to the existing body of knowledge, thus expanding human understanding of complex systems.

Model prediction, while not always of interest for a specific analysis, can be considered an extension of model interpretation. Recall that effective prediction does not necessitate the determination of causal relationships, only the identification of meaningful relationships between predictors and outcomes such that information about predictors can be used to infer associated response values. Prediction typically involves use of a model equation to determine expected values, and also acknowledgment of additional uncertainty associated with new, unobserved data. Prediction should not be confused with extrapolation, which is a dubious process of extending model relationships to unobserved ranges of predictor values. Instead prediction represents quantifiable expectations of future observations under similar conditions as those of the data used to build the model.

2.2.6 Effects and Predictive Model Differences

Effects models and predictive models share many model construction and fit assessment tests, and yet their respective purposes spawn differences in model structure and interpretation. These differences may be reflected in establishing fit statistics significance levels. The differences also manifest themselves through effects analysis interpretations.

A major difference between an effects model and a prediction model is that an effects model is a parsimonious rendering of the data whereas a predictive model's intent is to be robust against future data shocks. Effects models are used with the assumption that all possible configurations of a process or system are represented by the model construction data set. The goal is to find the predictors and their combinations that result in the most rigorous set of model fit statistics. The goal of an effects model, then, is to identify and quantify the most meaningful relationships between predictors and response. Often, no further use for the model is anticipated.

The purpose of a prediction model is forecasting future outcomes. A prediction model must be impervious to perturbations between the model construction data set to like-generated data. One technique to achieve a robust model is to retain marginally significant effects to account for latent or anemic samples in the data used for construction, but that may be unveiled or bolstered in forthcoming data. Another aspect of robust models is their ability to ignore incongruous samples within a data set. An ameliorative measure is the use of proxy variables to account for any additional variance to the incongruous samples. Prediction models, therefore, are intended to use as much information as possible to produce the most accurate representation of outcome values as is possible.

Essentially, effects models attempt a precise, parsimonious fit to a set of data whereas prediction models attempt to use all available information to estimate observed values, and allow overfitting to a data set.

2.3 Constant Variance Response Models

Normal linear regression models are appropriate for continuous responses that are assumed to have independent observations, to be normally distributed with equal variation, and that are linearly related to the regression parameters. A normal linear regression model can be written as an equation relating the outcome of interest to the predictors using regression parameters,

$$y_i = \beta_0 + \beta_1 x_{i1} + \cdots + \beta_p x_{ip} + \epsilon_i, \tag{2.1}$$

where y_i represents the response for observation i, the β_j represent regression parameters that describe the effects associated with each predictor, x_{ij} indicates the value of predictor j for observation i, and ϵ_i denotes the random error associated with the outcome for observation i. In this expression both y_i and x_{ij} are observed data, the β_j are parameters to be estimated, and the ϵ_i remain unknown errors. This model can also be written as a model for the mean of the response,

$$\mu_i = \beta_0 + \beta_1 x_{i1} + \cdots + \beta_p x_{ip}, \tag{2.2}$$

where μ_i indicates the mean for observation i, which no longer includes a random error because it no longer represents observed data. The parameter μ_i can be viewed much like the β_k, to be estimated using the model.

In order to predict or estimate mean values or perform hypothesis tests for predictors, it is necessary to obtain estimates for the regression parameters β_j. Such estimates, $\hat{\beta}_j$, are typically obtained by applying ordinary least squares to minimize the error sum of squares,

$$SSE = \sum_{i=1}^{n} \left(y_i - \hat{y}_i \right)^2. \tag{2.3}$$

The error sum of squares is the sum of the squared residuals, where in the formula $\hat{y}_i$ represents the estimated mean outcome for observation i, using the normal linear regression model. Solutions for the regression parameter estimates $\hat{\beta}_j$ can be written as a single, closed-form formula, and are appropriate regardless of the normality of the response of interest. However, the assumption of normality is important when performing subsequent hypothesis tests about predictors.

Normal linear regression models are often the first choice for modeling an outcome of interest. However, the normal linear model is not appropriate if any of the assumptions of independence, normality, constant variance, and linearity of the relationship are violated. There exists extensive literature describing how to adjust the normal linear regression model to account for violations of the assumptions; for example, adjusting the standard errors to perform more trustworthy hypothesis tests for parameters, or transforming both the response and predictors to make the outcome appear normal and the relationship linear. This post hoc adjustment approach is not the one we take in this text. Instead we take the approach of describing separate models that have been developed for different data situations and research interests, models that possibly do not include assumptions of normal responses, linear relationships, homogeneous variation, or independent outcomes. What follows is a brief description of these alternative models, covered more extensively throughout the text.

In Chapter 3, normal linear models are constructed for each of the four data sets described in Chapter 1 to establish a benchmark for comparing the efficacy of the models described below, and as well as show how efficacious the normal linear regression model is for these data sets.

2.4 Nonconstant Variance Response Models

It is common to encounter data situations in which the variation in the outcome of interest is not constant across the values of predictors. For example, outpatient rehabilitation clinics may record weekly patient volume, but the number of patients may fluctuate to a greater degree for clinics in urban areas as opposed to clinics in rural areas. As another example, education researchers may record the final grade point average (GPA) of a number of high school students, but find that GPA is much more consistent and shows less variability for students from families with higher household income levels. In both cases we see that the response variation changes according to one or more predictors, and normal linear regression models will produce erroneous results. Normal regression F-tests and t-tests are no longer trustworthy when the outcome variation is not constant.

Two classical approaches can be applied when heterogeneity in the response variance is encountered: transformations of the response, and weighted least squares. The goal of transforming the response values is to alter the scale of observation so that the changes in response variance do not appear as pronounced across values of predictors. Transformations are selected according to the nature of the response, or based on patterns observed in normal model residual plots.

Weighted least squares is an adjustment to the ordinary least squares estimation method that allows each response to be weighted according to a variable that represents the variation in that response. Specifically, responses with a relatively large weight are assigned greater influence in the least squares estimation, while responses with a relatively small weight are assigned lesser influence. In this way the effects of specific outcomes with large fluctuations are minimized to reflect the uncertainty inherent in data with large variation. Both parameter estimates and standard errors are adjusted, as least squares is performed by minimizing the weighted error sum of squares,

$$SSE_w = \sum_{i=1}^{n} \left(w_i(y_i - \hat{y}_i) \right)^2,$$ (2.4)

where y_i represents the outcome for observation i, $\hat{y}_i$ indicates the predicted outcome for observation i, and w_i indicates the associated response weight. Responses with larger weight represent larger values in the sum of squares, while responses with lesser weight represent smaller values. Appropriate weights are determined by the data analyst, and can include transformations of predictors or other variables relevant to the outcome. When the response of interest represents aggregated data, such as averages for groups, then the inverse of calculated group variances are often selected as appropriate weights. However, selection of weighting variables remains subjective and can differ by researcher.

Chapter 4 demonstrates nonconstant variance modeling using the School Survey on Crime and Safety data to predict the annual number of suspensions for insubordination, and uses the Fire-Climate Interactions data to model the decade averages of fire indicators using decade and region as predictors.

2.5 Discrete, Categorical Response Models

Recording outcomes as categorical variables is common in many disciplines. For example, student testing outcomes may be recorded simply as pass or fail, an industrial experiment may include results of the production of items as damaged or intact, or clinical health researchers may observe the presence or absence of a particular disease in patients. When the outcome of interest is recorded using two mutually exclusive categories, the response is said to be a binary response and normal linear regression approaches are not appropriate. Specifically, normal linear regression approaches are not suitable to handle the inherent changing variability associated with binary responses, the nonnormality of the observed binary data, and the typically nonlinear nature of the relationship between predictors and response. Instead data analysts typically turn to binary logistic regression.

Binary logistic regression is the most common modeling choice for outcomes that represent two categories. Using such a logistic regression model, predictors are used to estimate the probability associated with one of the two possible response categories of the outcome of interest. Therefore conclusions based on logistic regression models typically involve statements about the associated response probability, or about the response odds, which is the ratio of the probability of one response outcome to the probability of the other response outcome.

The logistic regression model can be thought of as a nonlinear model used to predict the probability of one response category,

$$\pi_i = \frac{\exp(\beta_0 + \beta_1 x_{i1} + \cdots + \beta_p x_{ip})}{1 + \exp(\beta_0 + \beta_1 x_{i1} + \cdots + \beta_p x_{ip})},$$ (2.5)

where π_i represents the probability of the outcome category of interest for observation i, exp() represents the natural exponential (or anti-log) function, the β_j represent regression parameters, and x_{ij} gives the value of predictor j for observation i. In this equation the x_{ij} are observed data, while the β_j must be estimated using the model, and are used to calculate π_i. The structure of the right-hand side of Eq. (2.5) ensures that predictions using any values of predictors and regression parameters will produce values between 0 and 1, which is appropriate for a probability. The logistic regression model is often re-written to look similar to a normal regression model,

$$\ln\left(\frac{\pi_i}{1 - \pi_i}\right) = \beta_0 + \beta_1 x_{i1} + \cdots + \beta_p x_{ip}, \qquad (2.6)$$

where ln() represents the natural logarithm function, the inverse of the natural exponential function. This form of the logistic regression model is representative of a class of models known as the generalized linear models.

The regression parameters in a logistic regression model cannot be estimated using ordinary least squares, as applied to normal linear regression models. Instead logistic regression parameter estimates are obtained through the method of maximum likelihood. Maximum likelihood is an estimation method that is based on an assumed distribution for the outcome of interest, in this case the Bernoulli (or binomial) distribution assumed for the binary outcome. With few exceptions, each assumed distribution has an associated likelihood function, which is maximized with respect to the parameters of interest to produce values of highest likelihood for the observed data. For the assumed Bernoulli distribution, the likelihood function takes the form

$$l(\beta_0, \ldots, \beta_p; y_1, \ldots, y_n, x_{11}, \ldots, x_{np}) = \prod_{i=1}^{n} \pi_i^{y_i}(1 - \pi_i)^{(1-y_i)}. \qquad (2.7)$$

The likelihood function includes information about the observed predictors (x_{ij}) and outcome values (y_i), as well as the relationship between the probability of the outcome category of interest (π_i) and the regression parameters (β_j). Results of maximum likelihood estimation are typically not obtained through a direct, closed-form formula, as is the case with normal linear regression. Instead iterative methods must be used to approximate solutions that maximize the likelihood function. Such iterative methods can often fail to converge because of properties of the data, which may mean that proper parameter estimates cannot be obtained for the logistic model.

Because of the nonlinear nature of the logistic regression model, we cannot make interpretations of parameters similarly to normal linear regression models. Most commonly, analysts interpret the anti-log (or natural exponential) function applied to each parameter, $\exp(\beta_j)$, which can be interpreted as the expected change in the odds of the outcome category of interest. Because the change in the odds represents a multiplicative change, this is referred to as the odds ratio.

When modeling outcomes that involve more than two response categories, ordinary binary logistic regression cannot be used. Instead extensions known as multinomial logistic regression models must be applied to account for the multi-categorical nature of the response. When considering such responses, there is no longer a single odds comparing one outcome category to the other. There are many outcome category comparisons that can be made, and therefore any multinomial logistic regression model involves multiple models providing predictions associated with different response probabilities.

For categorical responses that do not involve any inherent ordering, the nominal multinomial logistic regression model is appropriate. To apply this model, we arbitrarily select one of the outcome categories as a baseline category to which all other categories are compared and construct a binary logistic regression model for each comparison,

$$\ln\left(\frac{\pi_{ij}}{\pi_{iB}}\right) = \beta_0 + \beta_{1j}x_{i1} + \cdots + \beta_{pj}x_{ip}, \tag{2.8}$$

where π_{ij} represents the probability of observing category j for observation i, π_{iB} represents the associated probability for the reference response category, β_{kj} is the regression parameter associated with predictor k and response category j, and x_{ik} is the value of predictor k for observation i. In Eq. (2.8) the x_{ik} are observed values, while the β_{kj} are unknown and must be estimated through the model, and used to estimate each π_{ij}. For such nominal models, anti-logged (or natural-exponentiated) parameters are interpreted as odds ratios for comparing the probabilities of each outcome category to the baseline reference category.

For categorical responses that are inherently ordered, there are a number of ordinal multinomial logistic regression models that can be applied. Each ordinal model involves reducing to a comparison of two categories and applying a number of models that resemble ordinary binary logistic regression models. Typically, natural-exponentiated (or anti-logged) parameters are interpreted in terms of the odds of increasing from one response category to a higher category according to the inherent ordering. These models are discussed in more detail in Chapter 5.

Chapter 5 demonstrates the use of categorical response models. The probability of developing hypertension at any time after baseline in the Framingham Heart Study data is modeled using patient properties as predictors. The School Survey on Crime and Safety data are used to model the factors that impact school bullying, and the English Wikipedia Clickstream data are used to predict the probability of producing a redlink after redirection through Wikipedia.

2.6 Count Response Models

Count data are ubiquitous in empirical modeling. Such responses are recorded as the number of survivors of drug assessments and the number of emergency care facilities visitations. Epidemiologists may use count data to record the number of contagions and the number of deaths due to viral promulgation. The biological sciences use counts of bacteria in cultures to assess reproduction patterns and rates. The physical sciences use count data in studying numbers of sunspots and numbers of celestial objects within a specified field of view. Engineers use count data when recording the number of defects in manufacturing processes. School systems allocate resources based on the numbers of students in individual schools and within districts.

Normal linear regression approaches are not appropriate for count data as the mean and variance of counts are functionally related, whereas normal linear regression assumes the response variance is constant across mean values, implying independence of the mean and variance. Normal linear regression models are also a poor choice for count data because counts often show right-hand skew, violating the assumption of normality. Modeling such skewed data often requires nonlinear relationships between predictors and response, which is also incompatible with normal linear regression. Therefore normal linear regression approaches are not suitable to handle the inherent changing variability associated with count responses, the nonnormality of the observed count data, and the typically nonlinear nature of the relationship between predictors and response. Often, analysts transform a counts response using a square-root or natural-logarithm transformation to force an approximation of a Gaussian distribution. However, this shoehorn approach rarely succeeds as outliers often result. Fortunately, a class of models exists explicitly for count response models.

The count regression model is usually written to relate the mean or rate to predictors through a natural-exponential transformation, and may be represented as

$$\mu_i = \exp(\beta_0 + \beta_1 x_{i1} + \cdots + \beta_p x_{ip}). \tag{2.9}$$

In Eq. (2.9), μ_i is the mean count or rate of observation i, exp() is a transformation of the predictors and parameters to the same scale as the counts outcomes, $x_{i1}, x_{i2}, \ldots, x_{ip}$ are the observed values of the $1, \ldots, p$ predictors for observation i, and the coefficients to be estimated are β_0, β_1, β_2, $\ldots$, β_p.

The regression parameters in count regression models cannot be estimated using ordinary least squares as with the normal linear regression models. Rather, like logistic regression parameters, estimates are obtained using the method of maximum likelihood. Maximum likelihood estimation is based on an assumed distribution for the count response. The Poisson likelihood function, sometimes with adjustments, and the various negative binomial likelihood functions, are common choices for the distribution of count data. The associated likelihood function is maximized with respect to the parameters of interest to produce estimates of highest likelihood for the observed data. For the Poisson distribution, the likelihood function is

$$l(\beta_0, \ldots, \beta_p; y_1, \ldots, y_n, x_{11}, \ldots, x_{np}) = \prod_{i=1}^{n} \frac{\mu_i^{y_i} e^{\mu_i}}{y_i!}. \tag{2.10}$$

The likelihood function includes information about the observed predictors (x_{ij}) and count outcomes (y_i), as well as the relationship between the estimated mean count of the outcome (μ_i) and the regression parameters (β_j), as given in Eq. (2.9). Results of maximum likelihood estimation are usually calculated from an iterative method, which is used to approximate solutions that maximize the likelihood function. As mentioned previously, iterative methods may fail to converge because of properties of the data, in which case it is possible that proper parameter estimates cannot be obtained.

The nonlinear structure of the count regression model dictates that interpretations of parameters use the anti-log function applied to each parameter, $\exp(\beta_j)$, and is interpreted as the expected change of a rate ratio. These rate ratios may be interpreted as percentage changes (increasing or decreasing) in the counts. Also, these rate ratios may be expressed as multiplicative increases or decreases (e.g., a two-times increase in counts).

Count data are often considered to follow a Poisson probability distribution, and hence the name 'Poisson regression' is commonly used. An important characteristic of Poisson regression is the assumption of equi-dispersion; viz., the equality of the mean and the variance. The mean and variance are represented as

$$\text{Mean} = \mu, \qquad \text{Variance} = \mu. \tag{2.11}$$

However, data considered to follow a Poisson distribution may have variance greater than the mean. This is a condition referred to as overdispersion. Overdispersion may arise when there are violations in the distributional assumptions, such as when the counts are clustered, thereby violating the likelihood requirement of the independence of observations. It may cause standard errors of the estimates to be deflated or underestimated, i.e., a variable may appear to be a significant predictor when it is in fact not. If overdispersion is observed, then there are several corrective measures in common practice. Options include adjusting the standard errors by scaling, applying sandwich or robust standard errors, or bootstrapping standard errors for the model. However, using these methods will adjust only the standard errors and not the regression coefficients, β_k. Situations exist, however, in which the coefficients may be affected by overdispersion.

Overdispersion can be measured and often accounted for in Poisson regression, resulting in a so-called quasi-Poisson distribution that includes a dispersion statistic with a value greater than one, represented as

$$\text{Mean} = \mu, \qquad \text{Variance} = \mu + \phi\mu, \tag{2.12}$$

where ϕ is a nonnegative multiplier forcing the variance to be larger than the mean. When $\phi \neq 0$, a modification is made to the Poisson regression model in which the standard errors of the predictor parameters are adjusted by an amount proportional to the value of ϕ. The values of the predictor parameters do not change from those if $\phi = 0$. However, the quasi-Poisson model imposes the assumption that the variance must be a linear function of μ, and it is reliable only for relatively small amounts of overdispersion uniformly spread throughout the response data. Other options for accounting for overdispersion include the use of various negative binomial distributions for the count regression model. The negative binomial distribution allows the variance to be a quadratic function of the mean, making it functionally different from the quasi-Poisson distribution.

The definitions of both the Poisson and negative binomial models mandate an expected number of zeros in the data depending on each distribution's mean and variance parameters. If the count data have more than the expected number of zeros, we say the data have excess-zeros and should consider either a zero-inflated model or a hurdle model using adjustments of either the Poisson or negative binomial distribution. There are excess-zero versions for count distributions such as the inverse Gaussian, but we shall consider only the Poisson and negative binomial models in this handbook.

For counts with an inflated number of zeros, the zeros and counts greater than or equal to zero are modeled from two overlapping processes: a binary process to represent the presence of zeros, and a count process such as a negative binomial distribution to represent counts, possibly including zeros. The binary process is modeled using logistic regression, while the overlapping counts process is modeled using any appropriate counts regression model. This approach of two overlapping processes is referred to as a zero-inflated model. Because the processes overlap, zero-inflated models include the assumption that observed zeros can occur from either process. For example, in a model of school attendance, students may have zero absences because of good health (as part of the count process), but individuals may have no absences because of sanctioned extracurricular activities (as part of the binary process). The two components of an excess-zero model can be written as follows:

$$\pi_i = \frac{\exp(\gamma_0 + \gamma_1 x_{i1} + \cdots + \gamma_p x_{ip})}{1 + \exp(\gamma_0 + \gamma_1 x_{i1} + \cdots + \gamma_p x_{ip})},$$

$$\mu_i = \exp(\beta_0 + \beta_1 x_{i1} + \cdots + \beta_p x_{ip}). \tag{2.13}$$

In Eq. (2.13) the x_{ij} represent observed values of predictors, which need not be the same between the two components of the model, while the γ_j and β_j represent the unknown coefficients for the logistic and count components of the model, respectively, which are used to estimate π_i and μ_i.

A zero-deflated count model has the expected number of zeros less than that of a Poisson or negative binomial distribution. As with the excess-zero count models, we may consider having the zeros generated by a process different than that which generates the nonzero counts; viz., we may model the zeros and counts greater than zero from the overlapping binary process and a specified counts process such as the negative binomial distribution.

An alternate model for counts with inflated or deflated numbers of zeros is to model the zeros and counts greater than zero as two separate process rather than a mixture of these two processes. We use a binary process to represent the presence of zeros, and model the binary process with a logistic model just as with the zero-inflated (deflated) model. We then "hurdle" the zeros and model the nonzero counts with, say, a zero-truncated negative binomial model. The difference between this hurdle model and the zero-inflated model is that the truncated negative binomial model excludes the counts of zero rather than including them as a mixture as in the zero-inflated model. Hence, we are completely separating the zero and nonzero counts. The hurdle threshold may be a count higher than zero, e.g., all counts greater than 1. Hurdle models may be used with truncated count models in which the truncation is on the left, on the right, or both on the left and the right. Because the processes of zeros and positive counts do not overlap, the hurdle models include the assumption that the sub-population producing zero counts is distinct from the sub-population producing positive counts. For example, in a model of the number of nights spent in an astronomical observatory in a given week, observers will report zero nights due to cloud cover, while observers who experienced at least one night with good seeing conditions will report nonzero counts of nights spent in the observatory. These models are discussed in greater depth in Chapter 6.

Chapter 6 uses the School Survey on Crime and Safety data to model annual suspensions using various school characteristics. The Fire-Climate Interactions data decadal tree ring counts indicating the number of fires are modeled with several count models to demonstrate the efficacy of these models for the fire data. Also, the English Wikipedia Clickstream data are used to model its pairings counts versus the link type and the previous location type.

2.7 Time-to-Event Response Models

Time-to-event (TTE) data, also known as either survival data or as failure data, use a set of modeling methods designed to assess the time until some specified event occurs. Often studied events involve times to a cure, a failure, a departure, or a relapse. An example is a test for whether the survival is longer for a control group over a treated group for a specified disease therapy. An engineer may wish to know whether one relay switch is less likely to fail than another. Business and industry use the term failure analysis to evaluate the time for a part (relay switch, say) or a product to fail whereas medical investigators typically study times of survival after the occurrence of some event of interest.

The dependent variable, time, often is unknown for some individuals within a given study. It is common for a subject or product still in a study to have survived through the study period, or in other words to not have failed during an experiment. In medical studies, researchers must manage the situation where subjects withdraw from a study before the study is complete, and are thereby unavailable for follow-up analysis. Analogously, in industry, a part may be removed during scheduled maintenance even though the part has not failed. In such situations, these types of data are known as censored data. There are three types of censored data in time-to-event analyses. Left-censored observations refer to those whose event occurred prior to recording data, such as the death of a participant in a drug trial prior to the beginning of the study. Right-censored responses refer to those who do not show the event during the period of observation, such as a machine part that does not fail during the experiment. Interval-censored data refer to the situation in which the event is known to have occurred between two times, but the exact time of event remains unknown, which is common in studies with scheduled maintenance or scheduled medical check-ups.

Normal linear regression techniques are not appropriate for time-to-event data, as time values are not independent, they are typically skewed to the right, and are associated with predictors through nonlinear relationships. In addition, normal linear regression is not designed to handle the issues of censoring and the partial information associated with different types of censored observations.

Different procedures are used in survival analysis depending on the questions of interest. Tables and graphs describe and depict the survival rates of subjects, products, or events. As the rate is a function of time, the plot is often called a survival function. A useful plotting method for failure analysis is the mean cumulative function (MCF). The MCF is a nonparametric method that allows us to track cumulative events in time, and the first derivative (first difference) is the rate of change of events. There are several other linear modeling techniques to examine the effect of survival by event influencers.

A popular model for survival data analysis is the Cox regression model. A reason for its popularity is that it does not require using a probability distribution function to model the survival times. The Cox regression model has many implementations to allow for a variety predictor-to-predictor relationships. The Cox proportional hazards model is one such implementation, in which the hazard or rate at which individuals are expected to experience the event of interest is modeled. The Cox proportional hazards model has the form

$$\frac{h_i(t)}{h_j(t)} = \exp\left[\beta_1(x_{i1} - x_{j1}) + \cdots + \beta_p(x_{ip} - x_{jp})\right], \qquad (2.14)$$

in which the ratio of $h_i(t)$ to $h_j(t)$ is the hazard proportion of subject i to subject j as a function of the (possibly) time-dependent predictors x_{i1}, x_{i2}, $\ldots$, x_{ip} through the estimated parameters β_1, β_2, $\ldots$, β_p via the exponential function. This hazard ratio is assumed to be constant over time (t), suggesting that the survival behavior of any two subjects will plot as parallel curves. In this expression the x_{ik} are observed predictor values, while the β_k are unknown and must be estimated to produce values of the hazard ratio. Time-dependent extensions of the Cox model exist, allowing the hazard ratios to change over time. Cox regression, as opposed to the Cox proportional hazards model, accounts for hazard ratio changes with time.

Because of the nonlinear nature of the Cox proportional hazards model, we interpret the parameter estimates just as we do with logistic regression: the anti-log (or natural exponential) function is applied to each parameter, $\exp(\beta_k)$, which then is interpreted as the expected change or rate of occurrence of the event after time t. Because the change in this rate of nonsurvival represents a multiplicative change, this is referred to as the hazard ratio.

For life data analysis it is assumed that events are independent and identically distributed, but there are many cases for which this assumption does not hold. An example of these data, called recurrent events, is where the analyst models the number of occurrences of events over time rather than the length of time prior to the first event. An examination of recurrent event data is beyond the scope of this handbook.

Chapter 7 uses the Framingham Heart Study data to model the survival time from diagnosis of hypertension without predictors using life tables, and then with the Cox models to examine the influence of selected predictors.

2.8 Longitudinal Response Models

It is very common for multiple observations to be made using the same individuals over multiple times of observation. When the same outcomes are recorded from the same individuals many times, the data are referred to as panel longitudinal data, or often simply as longitudinal data or panel data. For example, school districts may track standardized exam grades for all students for each year of high school. Market analysts for an online shopping site can record the number of visits to a specific website each day for a month. A physician can observe and record indicators of dementia each time a patient visits her office. In all cases, the same individuals are followed over time, and the same outcome of interest is recorded at multiple moments of observation.

When analyzing panel longitudinal data, we cannot apply standard linear regression methods because panel data violate the assumption of independent responses. Responses from panel data are typically positively correlated, which is referred to as autocorrelation in the outcome. If this autocorrelation is ignored and standard normal regression techniques are applied, repeated associations observed for the same individual would be viewed as associations seen in multiple individuals, which would inflate the significance associated with the relationship. Instead we should apply longitudinal modeling techniques. Such techniques allow us to make inferences about growth or change over time, and allow us to evaluate associations or relationships among variables while accounting for the autocorrelation in the outcome of interest.

Longitudinal regression models can be broadly classified into one of two categories: marginal longitudinal models and conditional longitudinal models. Marginal longitudinal models are associated with population-averaged interpretations, which generally provide parameters that describe the expected change in the mean response between two populations that are similar but differ according to the associated predictor of interest. Population-averaged interpretations are equivalent to those made using standard regression techniques, but when we apply marginal longitudinal models we have accounted for the autocorrelation in the outcome. Conditional longitudinal models, on the other hand, are associated with subject-specific interpretations, which generally provide parameters that describe the expected effect on the mean response for an individual as the associated predictors of interest change over time. Subject-specific interpretations allow for dynamic statements about growth or change over time. Transition models, often considered as a third type of longitudinal model, are conditional models in which prior response values are treated as predictors in the model.

The most common method of marginal longitudinal estimation is the generalized estimating equations (GEE). Estimating parameters using the GEE involves two main steps: specifying a regression equation and selecting a working correlation structure. The regression equation is constructed similarly to cross-sectional regression functions, as predictors are selected and related to the mean of the outcome in a meaningful way,

$$\mu_{it} = f(\beta_0 + \beta_1 x_{it1} + \cdots + \beta_p x_{itp}), \qquad (2.15)$$

where μ_{it} indicates the mean outcome for subject i at time t, β_j are the regression parameters, x_{itj} is the value of predictor j for subject i at time t, and $f()$ indicates a function relating the mean to the predictors, such as the natural exponential function for count regression models. In this expression the x_{itj} are observed data, while the β_j are unknown and estimated through the model to produce values of μ_{it}. Data analysts choose the working correlation structure to represent the expected nature of the autocorrelation in the outcome of interest. This provides a mathematical description of the strength of the associations between responses for the same individual at different times. Given this information, parameter estimates are obtained by solving the GEE

$$\sum_{i=1}^{n} \left(\frac{\partial \mu_{it}}{\partial \beta_j} \right) \left(A_{it}^{1/2} R_{it} A_{it}^{1/2} \right)^{-1} (y_{it} - \mu_{it}) = 0, \qquad (2.16)$$

where y_{it} represents the outcome for subject i at time t, A_{it} describes the variation of said response, and R_{it} indicates the working correlation structure. Solutions are obtained by iteratively identifying the values of the β_j that result in a sum as close to zero as possible, usually determined a priori by the analyst or software.

Conditional longitudinal models are constructed in three main steps: in the first step, the sources of variation in the response, such as random variation or variation among a population of individuals who are measured repeatedly are identified. In the second step, a regression model, whose effects are conditioned on accounting for all of the sources of random variation identified, is constructed. In the third step, which components of the regression model are directly affected by the sources of variation identified in the first step are determined.

The simplest conditional longitudinal model includes a single source of variation in addition to the random unexplained variation, and is sometimes referred to as the random-intercept model. In this model the individuals of the panel longitudinal study are identified as a source of random variation, and only the intercept in the regression model is assumed to vary randomly among the population of individuals. The model can be written as a mixed-effects model,

$$\mu_{it} = f(\beta_0 + \beta_1 x_{it1} + \cdots + \beta_p x_{itp} + u_i),$$

where all terms are identical to those of the marginal longitudinal regression model given in Eq. (2.15), except μ_{it} represents the response mean accounting for the contribution of the random effect u_i, which is a normal random variable assumed to capture random fluctuations in the baseline outcome value among the population of individuals. Similarly to the β_j, the u_i are unknown effects that must be predicted through the model. It is this additional term u_i that accounts for the autocorrelation in the response, and imposes the assumption that all outcome values for an individual are correlated equally. More complicated conditional longitudinal models involve interactions between predictors and additional random effects, which are discussed in Chapter 8.

Parameters of conditional longitudinal regression models can be estimated in many ways, including maximum likelihood estimation, pseudo-likelihood estimation, h-likelihood estimation, and quasi-likelihood estimation, among others. Most methods are variations on the maximum likelihood process, in which a function describing the expected distribution of responses is maximized with respect to model parameters to identify values associated with the highest likelihood for the given data. All estimation methods are iterative, and thus suffer from some of the convergence concerns associated with likelihood estimation for other types of models. In fact, we have seen a greater prevalence of convergence issues with longitudinal models involving noncontinuous outcomes than with most other common models.

Chapter 8 demonstrates the use of panel longitudinal modeling on the Framingham Heart Study data and on the Fire-Climate Interactions data. The FHS data are used to model the probability of hypertension over the three waves of data collection. The Fire-Climate Interactions data will be modeled to predict the typical number of fire indications recorded over each decade while controlling for differences between regions and for variation among sites at different regions.

2.9 Structural Equation Modeling

Structural equation modeling (SEM) comprises statistical methods that allow examination of relationships among variables, particularly relationships with variables that cannot be measured directly. SEM can be viewed as the combination of factor analysis and multiple regression. Hypothesized variable relationships are schematically represented by path diagrams to represent the hypothesized set of connections suggesting an appropriate model. SEM analysis requires a model specification based on a known hypothesis. Model path diagrams are used to show the relationships of the variables. The effects and fit are estimated and evaluated to ascertain the strength of the relationships among the variables. One example of structural equation models is relating a Gaussian response to a set of predictor variables, i.e., normal linear regression. Another example is the influence of one variable on another variable through the effects of other variables that can be measured, such as investigating whether income is affected directly by type of trades training while the trades training type affects income indirectly through spatial acuity. A final example is the well-known problem of determining the level of intelligence in humans. Intelligence is not measured directly and is determined from measurements such as verbal comprehension, working memory index, perceptual organization index, and processing speed index. Correlations of these variables' measurements are examined for variances that indicate the measurement levels as outcomes of intelligence.

The seemingly unrelated regressions (SUR) model may be considered a special case of regression models. SUR models estimate parameters in a system of linear regressions that are interdependent. There is interdependence among the predictors across the system of models; the relationships between predictors are assumed to be endogenous. The unexplained response errors of a set of related regressions are often correlated, and this correlation structure informs the level of regression equation interdependence. The estimators from least squares are still valid if the responses are shown to be multivariate normal, or a large sample is available. These are not guaranteed, however, so EDA and model diagnostics are essential.

As an example of SUR, consider two companies, Company A and Company B, and data for these companies over the same period for capital stock prices, outstanding shares values, and gross investments. The two companies depend on each other's products for their own product manufacturing. The models are the gross investment of each company as a function of their respective outstanding shares values and capital stock prices. Thus, there are two regression models that can be analyzed separately, or as a linear system of equations. SUR analysis includes an examination of the two models' residual correlations to measure the amount of endogenous association. Significant residual correlations suggest the gross investment for Company A is highly dependent upon the gross investment of Company B, and vice versa.

SEM is based on the hypothesized model's variance-covariance structure associated with the variables representing the system of interest. An approach proposed by Bentler and Bonett (1980) has every variable, either measured or not measured, in a dependency role or in an influencing (loosely, predictive) role, intimating a regression-type relationship, as follows:

$$\zeta_i = \beta_{ij}\zeta_i + \gamma_{ik}\eta_k, \qquad (2.17)$$

where ζ_i is the ith response value, β_{ij} is the regression coefficient relating the ith and jth response values, γ_{ik} is a coefficient associating the ith response value to the kth influencer value, and η_k is the value of the kth influencer. In Eq. (2.17) both ζ_i and η_k are observed values, while β_{ij} and γ_{ik} are unknown effects to be estimated. An interesting aspect of SEM is that the responses can influence themselves, and hence the reason for ζ_i appearing on both sides of the equation.

Only the influencer variables η_k covary with, say, η_l, to form the covariance value ψ_{kl}. As such, the model parameters to be estimated are the β_{ij}, γ_{ik}, and ψ_{kl} from the data values ζ_i and η_k.

Chapter 9 demonstrates the use of SEM latent variable modeling on the Framingham Heart Study data to ascertain if stress influences the measured levels of hypertension and total cholesterol, by sex and age. The School Survey on Crime and Safety are used to examine the influence of school climate as measured by variables such as levels of crime and use of tiplines, as well as the influence on academic achievement as measured by test-taking ability and English-language competency.

2.10 Effect Size

It has become common for data analysts in many disciplines to report effect sizes with all effects analyses. Briefly, effect sizes represent standardized measures of the magnitude of an association between variables. Some effect size calculations rely on standardized differences between group averages, some utilize ratios of sums of squares to approximate explained variation, and others use noncentrality parameters from associated hypothesis tests.

The literature on effect sizes for linear modeling is quite extensive, and covers most cases of application of the normal linear model. However, there is a dearth of literature on effect sizes for many extensions of the linear model discussed in this handbook. In addition, available effect size measures for nonlinear models differ dramatically from those for linear models. For example, for logistic regression models it is often recommended to use odds ratios as estimates of effect size. However, odds ratios are values of multiplicative change whose meaning depends heavily on whether the value is greater or less than 1, and they bear little resemblance to the linear regression effect sizes based on ratios of sums of squares that are restricted to the range of 0 to 1. Therefore effect sizes have inherently different meanings and interpretations depending on the nature of the outcome of interest and the model applied. In general we believe that effect sizes have not been thoroughly investigated for nonlinear models, and recommend relying on parameter interpretations based on model coefficients as described in subsequent chapters.

2.11 Model Fit Measures

The purpose of model fit assessments is to provide measures of how well a model represents the characteristics of the data used to construct the model. The measures assess fit attributes such as the amount of variability accounted for by the model; whether the residual variance is homogeneous and whether residual values are independent of each other; and how well the model fitted values match the model generating data values. The various model types share some of the same model assessment measures, while some measures are specific to a particular model. Below are some of the more common and important model fit measures. The model fit measures are described in the context of two broad classifications: global measures of fit, and residual analyses.

2.11.1 Measures of Fit

In order to evaluate the overall fit of a model to data, analysts often turn to single measures that capture the quality of the fit. Such statistics are intended to provide a single value to give the data analyst a general idea of the adequacy of a model's representation of the data. These values are often used to compare modeling options, but some values have a fixed, absolute range to evaluate fit without comparison to competing models.

Normal Linear Models

For the normal linear model, the most common single statistic used to assess overall model adequacy is the coefficient of determination, commonly symbolized by R^2. (The adjusted coefficient of determination, or R_a^2, includes a penalty associated with the number of predictors.) The coefficient of determination is simply the proportion of variation in the outcome that is attributable to the model, and is restricted to a proportion between 0 and 1.

While R^2 is a useful descriptive statistic, we have found that many researchers rely on R^2 almost exclusively to evaluate normal linear model fit. We suspect part of the issue with R^2 is that the value falls within a small, fixed range, so that data analysts strive for models with values of R^2 as close to 1 as possible. However, it is possible to have meaningful models with extremely small values of R^2. We recommend the use of many statistics to evaluate fit, and in particular for normal linear models we suggest extensive use of residual analyses to assess model adequacy.

Logistic and Count Models

Both logistic regression and counts regression models fall into the class of generalized linear models, and therefore have similar fit measures. Because parameters for such models are not estimated using the least squares criterion, no value of R^2 is available. A number of "pseudo" R^2 measures have been developed for such models, including calculations of differences in log-likelihoods, for example. However, we do not recommend using such fit measures because of the lack of clarity in their interpretations and also because the values tend to be very small on the scale from 0 to 1.

The most common general fit statistics for logistic and count regression models are the Pearson χ^2 and the model deviance. The Pearson χ^2 statistic is a sum of squared residuals, scaled by a measure of variation,

$$X^2 = \sum_{i=1}^{N} \frac{(y_i - \hat{y}_i)^2}{V(\hat{y}_i)}, \tag{2.18}$$

where N is the number of observations, y_i is an individual observation, $\hat{y}_i$ is a predicted value, and $V()$ represents the relationship between the mean and the variance for the data type under consideration. Under certain conditions the Pearson statistic is distributed as a χ^2. Therefore a commonly used measure of fit is the ratio of X^2 to its degrees of freedom; a ratio that is much larger than 1 is taken as evidence of poor model fit. Similarly the model deviance is defined as a scaled difference between the logarithm of the likelihood function associated with the data type,

$$\hat{D} = -2\left(\ln(L_{\text{model}}) - \ln(L_{\text{data}})\right), \tag{2.19}$$

where ln indicates the natural logarithm function, L_{model} indicates the likelihood function evaluated at predicted values from the model, and L_{data} represents the likelihood function evaluated at the observed data values. The deviance is intended to represent the distance between the model and the data, where distance is measured in terms of the logarithm of the likelihood function. Under certain conditions, this statistic is distributed as a χ^2, and so the ratio of deviance to degrees of freedom is used to assess model fit. Many statisticians have cautioned that the conditions for this statistic to be distributed as a χ^2 are frequently violated, and therefore this fit statistic should be used in combination with others.

Information criteria are often used to compare potential logistic or count regression models. All information criteria are designed as measures of the information "lost" from data to model. The most popular such measure is the Akaike Information Criterion (AIC), which is often approximated using the largest value of the likelihood and is scaled according to the number of predictors in the model,

$$\text{AIC} \approx -2\ln(L_{\text{model}}) + 2p, \qquad (2.20)$$

where p represents the number of model coefficients, and the notation for the likelihood is as before. Values of AIC will always be positive, and smaller values are preferred. Statisticians have developed "corrected" AIC to penalize for the number of parameters, and other information criteria exist, such as the Bayesian information criterion (BIC). The purpose of all information criteria is to compare models with different combinations of the same predictors. Information criteria are not intended to be absolute indicators of fit, as any positive number can be produced for an adequate model.

Time-to-Event

Measures of model fit for Cox regression models are the χ^2 statistic, model deviance, the log rank score test, and the Akaike information criterion. Each of these statistics are described above in the logistic count model section.

A commonly used statistic for Cox regression is the score test. The score test is a test of whether a model coefficient is equal to a specified value. It is a most powerful test when the coefficient true value is close to the specified value.

Longitudinal Models

Longitudinal regression models have analogous fit measures such as the model deviance, the Pearson χ^2 statistic, and information criteria such as AIC. For estimation methods that do not rely on a full likelihood function, such as the generalized estimating equations, an alternative to AIC has been developed, called the quasi-likelihood under the independence model criterion (QIC). QIC can be used similarly to the AIC but is appropriate for a broader class of models.

Structural Equation Modeling

A good fit for a SEM suggests the hypothesis is plausible, but is not a confirmation that the specified hypothesis is correct. The statistics most often used to assess model fit are the χ^2 test, the root mean square error of approximation (RMSEA), the comparative fit index (CFI), and the standardized root mean square residual (SRMR). Several other fit statistics are available to assess the aptness of a model, and several should be considered for model fit. Each of these fit measures is described more fully in Chapter 9.

2.11.2 Residual Analyses

Analysts make use of residuals to evaluate the fit of all models. Briefly, residuals are thought to be representative of aspects of the response that were not captured by the model in question. While the global fit statistics of the previous section provide information about the fit of the model to the data, residuals provide information about the fit with respect to *individual observations*. Residuals can be defined in many ways depending on the model of interest, including raw residuals, studentized residuals, Pearson residuals, deviance residuals, marginal residuals, various forms of conditional residuals, and others specific to the types of models.

Residuals are constructed such that their distribution provides information about the quality of model fit. Residuals are usually assumed to be normally distributed, although nonlinear models often do not meet the requirements for derived residuals to be assumed normal. Nonetheless, normal Q-Q plots and tests of normality are commonly produced using residuals to assess normality assumptions or general model fit. Large absolute magnitudes of residuals can be used as evidence of poor model fit, indicating predicted values far from associated observed values.

It is also common to plot residuals versus fitted values or versus model predictors. Patterns in such plots can indicate an inappropriate model relationship, omitted predictors, or violations of response variance assumptions. Such residual plots are almost always used for linear regression model assessments.

Normal Linear Models

For linear regression models, such as normal linear regression and weighted least squares regression, fit diagnostics are well-developed. Residuals are defined simply as the difference between observed and predicted values. The assumption of normality is usually assessed by applying normality tests to residuals and creating associated Q-Q plots using residuals. Homogeneity of variance is evaluated using plots of residuals versus predicted values, in which changes in vertical fluctuation of values indicates violations of the assumption. The assumption of a linear relationship can be assessed by investigating the same residual plots, in which any visible pattern can suggest the need for a different model relationship. It is very common for these three assumptions to be violated together, meaning that if one fails the others often fail as well.

Logistic and Count Models

For generalized linear models, both Pearson and deviance residuals have been developed. Pearson residuals are calculated using the expected value and variation of each response,

$$r_i = \frac{y_i - \hat{y}_i}{\sqrt{V(\hat{y}_i)}}, \tag{2.21}$$

where all terms are defined as in the previous section. Notice Pearson residuals represent the individual observations' contributions to the general Pearson χ^2 statistic, and under certain conditions can be assumed normal. Deviance residuals are calculated as independent components' contributions to the model deviance, although the exact form depends on the type of data under consideration. Deviance residuals are often standardized, and under certain assumptions can be assumed normal. For both Pearson and deviance residuals, values large in absolute magnitude indicate observations for which the model fits poorly.

Pearson and deviance residuals can be plotted, but in many cases the plots can be misleading when used for linear regression models. For example, for binary logistic regression models, a plot of residuals versus predicted values will always show two bands of values, one for observations with a value of 0 and the other for observations with a value of 1. Similarly, such residual plots for counts regression models often show the "funnel" shape indicating nonconstant variation. While this pattern is a cause for concern with linear regression models, for count regression models it is expected. Fit statistics specific to logistic and count regression models are described in subsequent chapters.

Time-to-Event Models

Time-to-event models treat data differently than traditional linear or generalized linear models, and consequently traditional residuals such as deviance or Pearson residuals are not appropriate. Specifically, for time-to-event data any appropriate residuals must take into account the possibility of censoring, and the possibility that predictors change over time.

The most commonly applied residuals for assessing time-to-event model adequacy are the Schoenfeld residuals. A set of Schoenfeld residuals is produced for each predictor x_{ik} in a time-to-event model,

$$r_{ik} = c_i \left(x_{ik} - \frac{\sum_{i'=1}^{n} x_{i'k} \exp(\beta_0 + \cdots + \beta_p x_{i'p})}{\sum_{i'=1}^{n} \exp(\beta_0 + \cdots + \beta_p x_{i'p})} \right), \qquad (2.22)$$

where x_{ik} indicates the value of predictor k for subject i, c_i indicates whether subject i has been censored, i' counts subjects with observations at the time under consideration, and the β_j are the model coefficients. The Schoenfeld residuals are intended to capture each individual's contributions to the log partial likelihood across time.

Schoenfeld residuals are often used to assess the assumption of proportional hazards in a Cox proportion hazards regression model. For each predictor, the Schoenfeld residuals can be plotted versus time; a relatively flat trend near 0 indicates that there is no time-dependent effect that has been ignored by the proportional hazards model. These residuals can also be used to construct a χ^2 test to evaluate the proportional hazards assumption. Such methods are discussed in greater detail in Chapter 7.

Longitudinal Model

Longitudinal regression models include both Pearson and deviance residuals. In addition, longitudinal regression models include residuals calculated using different scales of the model, and residuals calculated with and without including values associated with the random sources of variation in the model. Scales used for residual calculation include the mean scale and the linearized scale. Conditional residuals include predicted values associated with random sources of variation, while marginal residuals typically assume the average value of zero for all sources of random variation.

The conditional Pearson residuals on the mean scale would be calculated as differences involving the observed values,

$$r_{p,it} = \frac{y_{it} - \hat{y}_{it}}{V(\hat{y}_{it})}, \qquad (2.23)$$

where $\hat{y}_{it}$ represents the predicted value for subject i at time t including the effects of any random sources of variation, such as a random intercept. Marginal Pearson residuals on the mean scale are calculated similarly, except with all random sources of variation replaced with mean values, typically zeros. The conditional Pearson residuals on the linearized scale involve transformations of responses to the scale of predictors,

$$r_{p,l,it} = \frac{g(y_{it}) - g(\hat{y}_{it})}{V(\hat{y}_{it})}, \qquad (2.24)$$

where $g()$ indicates the link function, such as the logit function for logistic models or the natural logarithm function for count models. In this way $g(\hat{y}_{it})$ becomes the right-hand side of the model equation. The term $g(y_{it})$ is usually approximated using a Taylor series. In general we prefer using residuals on the mean scale, as mean-scale residuals are more closely associated with the scale of the original measurement. Deviance residuals can also be adjusted to be conditional or marginal, but rarely are used on the linearized scale. Generally, these residuals are used to identify poor-fitting observations by unusually large values, and under certain assumptions can also be assumed normal and used to assess model fit by evaluating normality as usual.

2.12 Summary

We opened the model-building chapter with a description of the role of models for understanding objects or processes, whether a model is empirical or mechanistic, whether a model is intended for effects analysis or prediction, and that well-posed and well-constructed mathematical models allow us to add to the body of knowledge.

We focus on empirical models in this handbook. Empirical models describe causal relationships only with studies designed specifically for this purpose. We consider those empirical models that describe only the influence or association or relationship between the predictors and the response, both of which are observed. Empirical models are used to understand the particulars how predictors influence responses, and in this text we refer to such models as effects models. However, empirical models often are used to predict outcomes for combinations of predictors not necessarily found in the data set used to construct the model. Also, predictive models may be used to forecast outcomes yet to occur. In either case, a model used for prediction or forecasting must be robust against the specific behaviors of the variables in the construction data set. Our experience with prediction and forecasting recommends that any parameters estimated by the modeling process retain as many digits in the estimates as practical. This retention of digits minimizes error propagation that can lead to inaccurate outcomes.

The model-building process was described as iterative in that the model-building process itself often uncovers unknown relationships that require a proper accounting. In this situation, an original model usually is revised to make the accounting.

The implication is we must decide if the constructed model is appropriate for the intended purpose, and has an adequate fit to the data. Most importantly, the model must provide information that permits insight into, and interpretation of, the nature of the process or study goals as revealed from the data.

The process for constructing models was described as a five-part operation. The first, exploratory data analysis, is used to characterize the data to help choose an appropriate model type, variable conditioning, and obstacles that may require special attention. Secondly, the model construction including parameter estimation methods was discussed. Upon completing a candidate model, several diagnostic methods were presented to ascertain model adequacy. If the model diagnostics are acceptable, then the effects are analyzed for response changes to predictor level settings. Finally, the model outcomes and predictions were described.

As this handbook is concerned with the analysis of non-Gaussian and correlated data, we introduced the model types and their application to the varieties of data in use or to be collected. The model descriptions began with the commonly used constant variance response models such as normal linear regression to establish a baseline for comparing the adequacies of the model types we apply. The model types applied are models for nonconstant response variance such as regressions using weighted least squares, models for discrete categorical responses, and then count response models. These descriptions were followed by discussions on time-to-event response models for survival and failure analysis, longitudinal models for describing outcomes gathered over time, and finally, we described models for which information on a variable is desired, but the variable itself is not measurable, requiring the use of proxy variables, for example, to understand its behavior.

In the chapters that follow, we expand on the introductions to the model types described in this chapter by applying each model to one or more of the data sets described in Chapter 1. These chapters demonstrate the use of the model types and compare them with other model types, particularly with the normal linear regression model. We thereby demonstrate the efficacy of correctly matched data types to models.

2.13 Further Reading

More references for each model are included in each associated chapter. Briefly, details on using the normal linear model can be found in Kutner et al. (2004). A more abstract treatment of statistical modeling, including nice intuition on hypothesis testing, is available in Snedecor and Cochran (1989).

The logistic and count models are treated in great generality in McCullagh and Nelder (1989) and also in Dobson (2002). The logistic model in particular is described very clearly in Hosmer et al. (2013) and also in Hilbe (2015). Count regression models are treated very understandably in Hilbe (2014).

An excellent general reference for all types of longitudinal data analysis can be found in Diggle et al. (2002). The generalized estimating equations are treated thoroughly in Hardin and Hilbe (2003).

A nice applied reference for time-to-event models can be found in Hosmer et al. (2008).

Much of the information necessary to become comfortable with various statistical software can be found on the World Wide Web. For R, see Crawley (2007) and Verzani (2014).

3

Constant Variance Response Models

3.1 Introduction

In this chapter we discuss the use of *normal multiple linear regression* models. The models are said to be *normal* because we assume the errors (and consequently the responses) are distributed normally. The term *multiple* refers using more than one predictor to account for the variation in the response variable. We say the models are *linear* in the parameters as described in Chapter 1. The prediction functions do not need to be linear, but the regression coefficients cannot be involved in any nonlinear expressions. *Regression* refers to the estimation method that regresses a data-fitted line toward the true line which rarely is achieved or known.

Recall that a model is useful only if the data and variables are appropriate to the subject matter. When this is the case, we may construct a model for which we expect the predictors to influence changes to the response.

Multiple predictors are important for describing changes in the response variable for two primary reasons: processes for modeling are rarely simple, and multiple numbers of predictors usually more thoroughly partition the variance exhibited in the response. Pairing individual predictors with the response to assess pairwise correlations averages out variation due to other omitted predictors, leaving the amount of unexplained variance large. Often the level of a response is better understood when two or more predictors interact. The level of one predictor may alter the correlative relationship with some other predictor on a response. Another reason for multiple predictors in a predictive regression model is it often better accommodates new data.

We examine the ability of normal multiple linear regression to produce viable models on each of the four data sets presented in Chapter 1. We subject each data set to normal multiple linear regression for two reasons: firstly, we have assumed that users of this handbook are familiar with basic statistics and normal multiple linear regression; and secondly, we use these analyses as a baseline, or reference, from which to compare the outcomes of the models presented in Chapter 2. These outcomes are presented in the chapters following this one.

3.2 School Survey on Crime and Safety

We are interested in using the School Survey on Crime and Safety to predict the number of annual suspensions using characteristics of a school, including the number of insubordinate students during the year of interest, the percent of students with limited English-language proficiency, a measure of the crime in the area of the school's location (low, moderate, or high), and indicators of whether the school requires students to wear uniforms, whether students pass through metal detectors when entering the school, whether the school has a tipline to report issues, whether counseling services are available for students, and whether the teachers have available training in discipline policies or in positive behavioral interventions. We included interaction terms to allow for the effects of training in discipline policies and positive behavioral interventions to change across area crime levels. The model can be written as follows:

$$
\begin{aligned}
\mu_{\text{Suspensions}} = {} & \beta_0 + \beta_1(\text{uniforms}) + \beta_2(\text{metal detectors}) \\
& + \beta_3(\text{tipline}) + \beta_4(\text{counseling}) + \beta_5(\text{moderate crime}) + \beta_6(\text{high crime}) \\
& + \beta_7(\text{discipline training}) + \beta_8(\text{behavioral training}) \\
& + \beta_9(\text{insubordination}) + \beta_{10}(\text{limited English}) \\
& + \beta_{11}(\text{discipline} \times \text{moderate crime}) + \beta_{12}(\text{discipline} \times \text{high crime}) \\
& + \beta_{13}(\text{behavioral} \times \text{moderate crime}) + \beta_{14}(\text{behavioral} \times \text{high crime}).
\end{aligned}
$$

$$(3.1)$$

We report the results of evaluating the model assumptions, but due to poor diagnostic measures we will omit interpretations of parameter estimates and discussions of significance. After applying a test of the homogeneity of variance to the model residuals, we find a test statistic of $X^2 = 199\,732.5$ with associated p-value $p < 0.001$, less than any reasonable significance level and suggesting strong evidence of heterogeneity in the residual variance. As further support of this conclusion, the scatter plot of residuals versus predicted values shown in the left panel of Figure 3.1 shows a clear "funnel" pattern in which the dispersion in the residuals clearly increases as the predicted values increase.

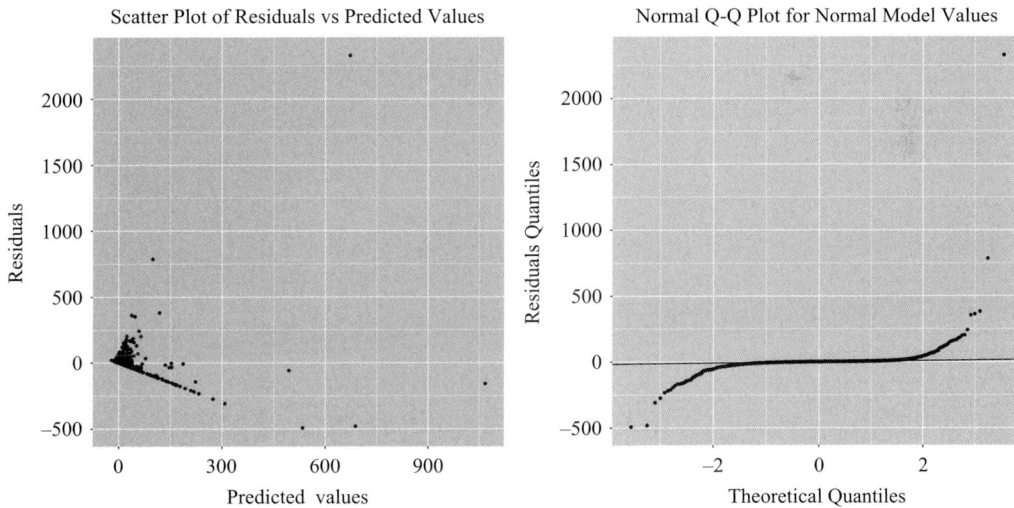

Figure 3.1 School Survey on Crime and Safety suspensions model, plot of residuals versus predicted values (left panel) and normal Q-Q plot of residuals (right panel).

To evaluate the assumption of normal residuals, we applied the Anderson-Darling test and found a test statistic of $A = 602.93$, with associated p-value $p < 0.001$, smaller than any common significance level. The normal probability plot shown in the right panel Figure 3.1 shows the "heavy tail" trend of the plot becoming vertical at the extreme left and right, suggesting potential outlier values that were not well captured by the model and leading to an inflation of the Type I error rate. The data provide clear evidence of nonnormality.

We used the linear model from Eq. (3.1) to make predictions of annual suspensions. Predictions for the number of annual suspensions range from -21.09 to $1\,068.00$, with $1\,101$ predictions less than 0. There is no reasonable way to interpret a prediction of a negative number of events during any time period, so these predictions clearly do not make any practical sense.

For example, for a school in an area with low crime with a required school uniform and discipline policy training available to teachers, but no metal detectors, tipline, counseling, or positive behavioral training, and with 50% of students with limited English-language proficiency and 5 insubordination events during the year of interest, the normal linear model produces a prediction of -6.679 suspensions during the school year. There is clear evidence that the normal linear model fits poorly, and it easily produces unrealistic predictions for the outcome of interest.

3.3 Framingham Heart Study

We have an interest in applying the Framingham Heart Study data to model the prevalence of hypertension using the predictors sex, diabetes, total cholesterol, age, and the number of cigarettes used per day. Hypertension is recorded numerically in the data set as either a 0 (no indication of hypertension) or a 1 (hypertension is present), so a normal linear model will fit a line between the 0s and 1s, essentially modeling the probability of the presence of hypertension. We included interactions to allow the effects of both the presence of diabetes and the number of cigarettes smoked per day to change between males and females. The model can be written

$$
\begin{aligned}
\pi_{\text{Hyp}} = {} & \beta_0 + \beta_1(\text{sex}) + \beta_2(\text{diabetes}) + \beta_3(\text{cholesterol}) + \beta_4(\text{age}) \\
& + \beta_5(\text{cigarettes}) + \beta_6(\text{sex} \times \text{diabetes}) + \beta_7(\text{sex} \times \text{cigarettes}),
\end{aligned}
\tag{3.2}
$$

where π_{Hyp} indicates the mean or probability of the presence of hypertension. We report the test results for model assumptions, but because of the low quality of the model fit we will omit interpretations of parameter estimates and discussions of significance. Applying a test of the homogeneity of variance to the model residuals, we find a test statistic of $X^2 = 360.24$ with associated p-value $p < 0.001$, significant at any commonly selected level. The data have provided clear evidence of heterogeneous variation. In addition, the plot in the left frame of Figure 3.2 shows a clear pattern, violating the normal linear regression model assumption of constant variation.

Next we applied the Anderson-Darling test of normality to the model residuals, which gives a test statistic of $A = 1\,315.9$ with associated p-value $p < 0.001$, significant at any typical preset level. In support of this conclusion, the normal probability plot shown in the right frame of Figure 3.2 does not resemble a straight, diagonal line, instead showing the "lazy S" shape that is common of residuals showing less variation than is typically

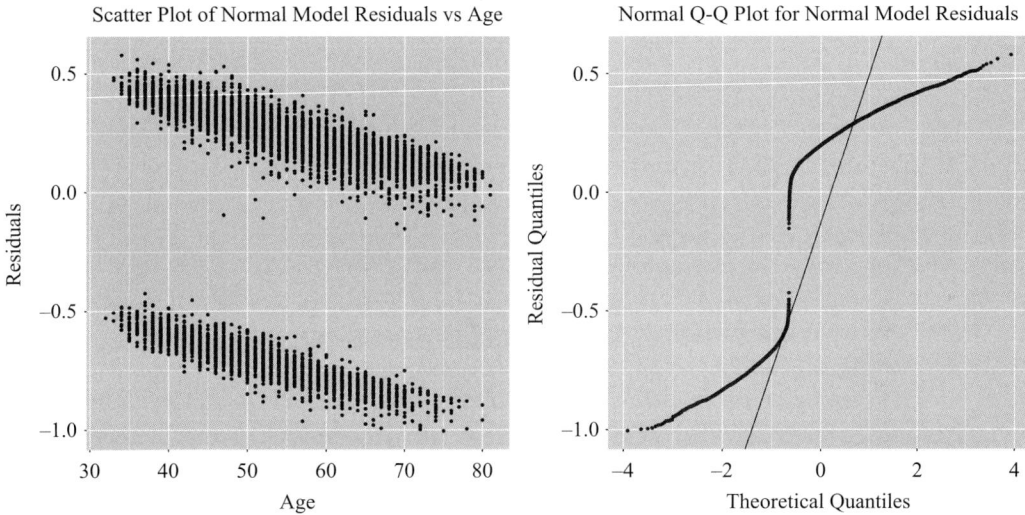

Figure 3.2 Framingham Heart Study hypertension model, plot of residuals versus the age predictor (left panel) and normal Q-Q plot of residuals (right panel).

expected of a normal distribution. The data have provided clear evidence of nonnormality of the residuals. Considering the source of the data, we know the Framingham Heart Study includes three waves of observations recorded for the same individuals, and therefore we know the data fail the assumption of independence required for the normal linear model.

We used the normal linear model to make predictions of hypertension. Predicted values for hypertension range from 0.4220 to 1.1540, with 65 cases predicted to be greater than 1. However, all values for hypertension in the data set are either 0 or 1, meaning that any of these predictions that are greater than 1 goes beyond the range of reasonable values for the outcome of interest. If we consider the interpretation of the mean of hypertension as a probability or proportion of individuals within the described population who should show evidence of hypertension, these predictions remain nonsensical.

As an example, consider a prediction of expected hypertension for an 80-year-old female with diabetes, reporting use of 80 cigarettes per day and showing a total cholesterol of 650, all values within the range observed for the data. The predicted value for hypertension is $\hat{\pi}_{\text{Hyp}} \approx 1.102$. A proportion of 1.102 would indicate that more than 100% of the individuals within the population described by the predictors would be expected to show evidence of hypertension. Clearly, the normal linear model provides poor fit and unrealistic predictions for hypertension.

3.4 Fire-Climate Interactions in the American West

We utilize a simple model of the counts of tree rings to detect fire scarring within the decades from 1130 through 2000 of the Fire-Climate data. The predictor variables chosen are decade and region, the location in which the sampled trees are found.

The region predictor level "IW" is used as the reference level from which the coefficients of "NC" and "PNW" levels are estimated. To avoid issues with multicollinearity, we center

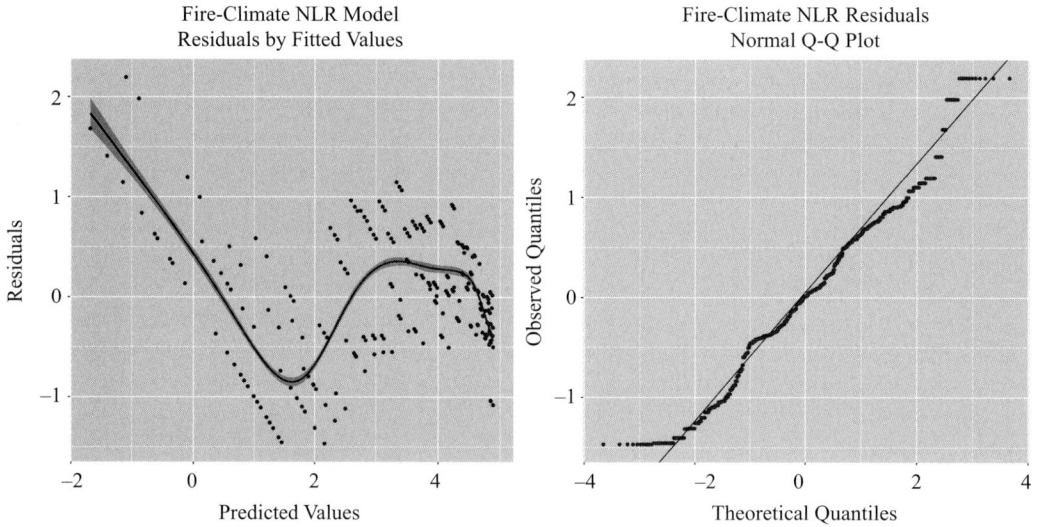

Figure 3.3 Fire-Climate Interactions data. The left-hand panel is the model residuals versus the model fitted values. The right-hand panel is a Q-Q plot of the residuals.

decade on its mean as "centered decade = decade - mean(decade)," where "mean(decade)" is the mean of "decade." The scatter plot of "decade" by "ln(ring count + 1)" is nonlinear, and we therefore test the quadratic form of "decade." Finally, to preserve model hierarchy, we include the two-way interaction of "decade" and the region levels. The normal linear regression model for the number of tree rings per decade is written

$$\mu_{\text{ln(ring count)}} = \beta_0 + \beta_1(\text{NC}) + \beta_2(\text{PNW}) + \beta_3(\text{centered decade})$$
$$+ \beta_{13}(\text{NC} \times (\text{centered decade})) + \beta_{23}(\text{PNW} \times (\text{centered decade})) \quad (3.3)$$
$$+ \beta_{33}(\text{centered decade})^2.$$

A test of the homogeneity of variance of the model residuals gives a chi-square statistic of $X^2 = 78.13337$ with associated p-value $p \ll 0.01$, less than any common significance level, evincing strong evidence of nonhomogeneous variance in the residuals. The scatter plot of the fitted values versus the predicted values shown in the left-hand panel of Figure 3.3 shows clear nonsymmetrical, almost sinusoidal pattern in which the dispersion in the residuals is increasing as the predicted values increase, thus depicting nonconstant residuals variance.

The Anderson-Darling test for Gaussian residuals gives a test statistic of $A = 13.772$, with associated p-value $p < 0.001$, smaller than any common significance level. The normal probability plot shown in the right-hand panel of Figure 3.3 shows light tails at the extreme left and right leading to an inflation of the Type I error rate. The residuals data provide clear evidence of nonnormality even with a log-transformed response.

The predictions of decadal fire-scarred tree ring counts based on the normal linear model gives values inconsistent with observed outcomes. The observed values range from 0 to 80 whereas the predicted values range from -1 to 60. In fact, the normal linear model predicts 32 negative values which are disallowed in count data.

An example of a negative prediction of expected counts of indications of fire is for the Pacific Northwest region in the decade from 1140 to 1149. These predictor values are within the range observed in the data. The predicted value for the number of fires is $\hat{\mu}_{\text{ln(ring count)}} \approx -0.729$. Clearly, a negative count is undefined for this problem, and suggests the normal linear model is inadequate for the fire-climate data.

Nonconstant variance, residuals with a non-Gaussian probability distribution, and unrealistic model predictions are clear indicators that using normal linear regression to model the fire-scarred tree ring counts is a poor choice for representing these data.

3.5 English Wikipedia Clickstream Data

We wish to predict the number of pairings of referrer and requested site using the English Wikipedia Clickstream data. The predictor variables for the number of pairings are the link type and the referring location. The response is the natural log of the pairings counts, which are truncated on the left at 10, eliminating counts of 0 through 9.

The reference level for the pair-type categorical variable is "link" from which the levels "other" and "redlink" parameters are estimated. The reference level of the "previous title" categorical variable is "Bing," from which the other previous title level parameters are estimated. The normal linear regression model is

$$
\begin{aligned}
\mu_{\text{ln(pair count)}} = {} & \beta_0 + \beta_1(\text{other(type)}) + \beta_2(\text{redlink}) + \beta_3(\text{empty}) + \beta_4(\text{Google}) \\
& + \beta_5(\text{Main\_Page}) + \beta_6(\text{other(previous title)}) \\
& + \beta_7(\text{Wikipedia}) + \beta_8(\text{Yahoo}).
\end{aligned}
\tag{3.4}
$$

The test of homogeneity of variance of the model residuals has a chi-square statistic of $X^2 = 144\,601.3$ with associated p-value $p < 0.001$, which is much less than the usual significance levels, and thus we have strong evidence of residuals with heterogeneous variance. Further, the scatter plot of the residuals versus the fitted values, shown in the left-hand panel of Figure 3.4, shows an increase in variation from small predicted values to large values, suggesting nonconstant residuals variance.

The Anderson-Darling test for normal residuals has a test statistic of $A = 1\,953$, with associated p-value $p < 0.001$, smaller than any reasonable significance level. The normal Q-Q probability plot shown in the right-hand panel of Figure 3.4 shows extremely heavy tails on the left and right of the plot leading to a highly inflated Type I error rate. The residuals data deviating from the Gaussian quantile line is clear evidence of the violation of normal linear model residuals distribution.

The fitted pairings counts range is from 20 to 55 and yet the observed pairings counts log values range from 10 to 95 738. Clearly, the model produces a mismatch between the two scales resulting in invalid predictions.

We see that the heteroscedastic residuals' variance, the non-Gaussian probability distribution of the residuals, and mismatch between the model predictions and the observations clearly demonstrate that normal linear regression is not an appropriate model of the clickstream count data.

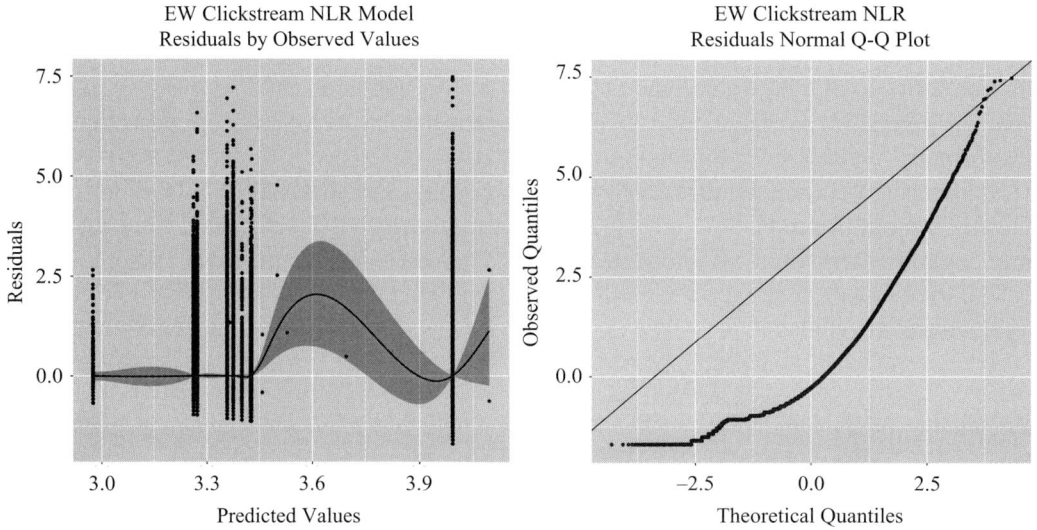

Figure 3.4 The left-hand panel is the model residuals versus the model fitted values. The right-hand panel is a Q-Q plot of the studentized residuals.

3.6 Summary

Traditional normal (Gaussian) multiple linear regression was used on each of the four data sets introduced in Chapter 1. The EDA of each data set suggested that normal linear regression as model choice was risky at best. We used normal linear regression as a reference model to compare with the other model types described in the remainder of this handbook. Often, analysts use normal linear regression as it is familiar, and some literature suggests that normal linear regression is robust against some deviations from the Gaussian assumption.

However, as the fit statistics and residuals analyses on the four data sets has shown, using normal linear regression as a model for these outcomes proved ineffective. Each model's residuals fail the requirement that they follow a normal or Gaussian distribution and that they have homogeneous variance across the fitted values. Perhaps even more importantly, the normal linear model produced predicted values for each data set that are completely unrealistic relative to the data themselves.

Normal linear regression, then, should be used only on those data that comply with the model's assumptions of straight-line relationships between the outcome and model parameters, and the normality, constant variation, and independence of residuals. The remaining chapters further address appropriate data set-to-model type matches.

3.7 Further Reading

A comprehensive treatment of the linear model is available in Kutner et al. (2004). Some nice texts explaining the use and application of the linear model are Freedman (2009), Glass and Hopkins (1995), and Agresti and Finlay (2008).

4

Nonconstant Variance Response Models

4.1 Heterogeneity in Response Variance

One of the assumptions of the normal regression model is the homogeneity of variance. That is, it is assumed that each response represents an outcome from a (normal) distribution with the same, constant variation, regardless of the values of associated predictors. However, this property of homoscedasticity is often violated by responses that show fluctuations in variation across values of predictors as was seen in Chapter 3 using the four data sets. In such cases researchers will still be interested in evaluating all of the same questions that can be answered using normal regression models, but must turn to alternative modeling techniques.

For example, education researchers may collect records of standardized test scores for students in multiple classrooms with an interest in comparing performance across different student demographics. However, in analyzing the data it may be seen that students in classes with more experienced instructors show more consistency in their performances, and therefore lower variation in scores than for students in classes with less experienced instructors. As another example, health policy researchers might be interested in connections between family income level and health and medical expenditures. When data are collected, data analysts may see that the health spending varies greatly for families with significant financial resources, but health spending is consistent for families less financial resources. In both examples the variability of the outcome of interest changes depending on different subject characteristics.

Within any normal regression model, inferential methods such as t-tests and F-tests are constructed under the assumption of homoscedasticity. Using this assumption, "pooled" estimates of variation are made using the data, including the random unexplained variation captured by the mean-squared error (MSE). If such inferential methods are applied to heteroscedastic data, the pooled estimates such as MSE can either underestimate or overestimate the true variation in the response, and therefore hypothesis-testing procedures will be unreliable, often in unpredictable ways. Therefore it is important for data analysts to be able to detect and correct for violations of the constant-variance assumption.

When data analysts encounter heteroscedasticity in an outcome of interest, one of two common remedies is often employed. The first is to transform the response of interest. The intention of applying a so-called variance-stabilizing transformation is to scale the outcome observations such that the variation appears to be constant across predictor values within the model. The second approach is to apply weights to the parameter estimates and standard errors. Such weights are selected to reflect changes in the outcome variation. Capturing these

57

fluctuations in variation through a weighting variable allows the changes in dispersion to be accounted for in parameter estimates and standard error estimates, correcting the bias that leaves hypothesis tests unreliable in the original model.

4.2 Detecting Heteroscedasticity

An important first step in dealing with nonconstant response variation in a regression model is identifying the problem of heteroscedasticity. It is common to employ both visual descriptive statistics along with inferential tests to evaluate the homoscedasticity assumption. For very small data sets, it is possible that visual displays will struggle to show patterns, and the additional information of hypothesis tests is important. For data sets with a large number of observations relative to the number of predictors, the power of hypothesis tests of the constant variance assumption will be very high, such that the data will be able to detect very small discrepancies from homoscedasticity. Therefore the test is overpowered, and p-values can suggest significant departures from the assumptions that are not practically important. The general trends shown in plots are then more valuable than the tests. Visual statistics and hypothesis tests should be used in conjunction whenever possible.

4.2.1 Descriptive Statistics

Most descriptive statistics used to evaluate the assumption of homogeneous variation involve residual plots. Using an otherwise appropriate normal linear model, residuals can be obtained and plotted on the vertical axis with model predicted values on the horizontal axis. The variation in the response is captured by the variation in the residuals within a linear model; it is the vertical range of values that reflects this variation in the residual plot. When this vertical range changes from left to right across the plot, there is evidence of heterogeneous variance. In addition to residual plots using the predicted response values on the horizontal, using individual predictor values on the horizontal can identify which specific predictors are associated with changing response variance.

4.2.2 Tests for Grouped Data

For predictors that represent classifications of observational units in the study, such as gender or education level, residual scatter plots are typically replaced by residual box plots to identify nonconstant variation. In such plots a separate box plot of residuals is constructed for each classification in the model of interest. Changes in the variation indicated by box plots suggest that the response suffers from heterosedasticity.

In addition to box plots, the assumption of constant variance can be assessed using appropriate tests. For outcomes grouped by a classification variable, Levene's test can be applied as a test of constant variance. Briefly, Levene's test uses the magnitude of response deviations around group means, $|Y_{ij} - \bar{Y}_{i.}|$, where Y_{ij} is the jth outcome in the ith group and $\bar{Y}_{i.}$ is the average of all of the responses for the ith group. Levene's test is based on a statistic that compares the variation of these deviations across groups to the variation of the deviations within groups. If there is greater variation across groups, it is taken as an indication that the variance differs across classifications in the model.

4.2.3 Tests for Continuous Predictors

For regression models involving continuous predictors, Levene's test is not appropriate. Instead options such as White's test and the Breusch-Pagan test should be applied. Briefly, both the Breusch-Pagan test and White's test involve building a new regression model using the squared residuals (from the original normal linear regression model) as the outcome of interest. The squared residuals from the original model are regressed against all of the original predictors. Because squared residuals are part of the calculation of variation, if any predictor shows significance in this secondary model, it is taken as evidence that the predictor is associated with changes in variation and the assumption of homogeneous variance is violated.

4.3 Variance-Stabilizing Transformations

When dealing with a model for which the homoscedasticity assumption fails, a traditional remedy is to apply a variance-stabilizing transformation to the response. A variance-stabilizing transformation is intended to expand or compress the scale of the response so that the observed changes in variation are no longer significant.

4.3.1 Selecting the Transformation

Variance-stabilizing transformations are typically selected somewhat informally. For certain types of data, common transformations may be applied. For response data that represent counts or frequencies, a logarithmic transformation ($\ln(y)$) can often be applied to account for the increasing variation that is typical of count data. A logarithmic transformation is often applied to generally right-skewed data. When applying a logarithmic transformation to a response, data analysts often apply a small "adjustment" to the values to avoid generating infinite values, such as using $\ln(y+0.001)$ as the transformation of choice. For responses that are proportions of two variables, the inverse-sine-square-root transformation ($\arcsin(y^{1/2})$) has been found to be effective in handling the decreasing variation at either endpoint of allowable proportions (usually 0 and 1).

Data analysts often consult residual plots to determine appropriate transformations. When the residuals are plotted against the predicted response values, the changes in vertical fluctuation in the plot give an indication of nonconstant variability in the data. When the residual plot shows a straight-line increase or decrease in variation across values of the predictor (the "funnel" shape), a square-root transformation ($y^{1/2}$) has been shown to be effective. For funnel shapes that do not appear to be linear, but are rather concave and show a growth or decay that is slower than linear, a logarithmic transformation is often applied. Finally, for residual plots that show an increase in variation followed by a comparable decrease in variation, the inverse-sine-square-root transformation is often successfully applied.

4.3.2 Model Diagnostics

After transforming the outcome variable to accommodate nonhomogeneous variation, a new regression model can be fitted using the transformed data as the new response. After fitting the model, it is important to once again evaluate the homogeneity of the variance.

Unfortunately, interpretations of the regression parameters often become challenging when dealing with transformed data. For example, when a logarithmic transformation is applied, any regression coefficient represents the expected change in the mean of the logarithm of the original response, for a unit increase of the associated predictor across populations. If the original response is the number of students who register for a class, the researcher is typically interested in answering questions about the mean number of students. Providing interpretations about the mean of the logarithm of the number of students may not be useful. Instead researchers often rely on simple explanations in terms of general "increases" or "decreases" in the expected response, and on predictor significance.

4.4 Weighted Least Squares

As an alternative to transforming the response of interest to account for heterogeneous variation, weighted least squares (WLS) can be applied in place of ordinary least squares (OLS) to estimate model parameters. In this case the model itself does not change; instead the method of obtaining parameter estimates and standard errors changes.

4.4.1 WLS Estimation

OLS estimation involves a process of minimizing the sum of the squared errors; the vertical distances between the proposed regression line and the observed responses are squared, summed, and then minimized to obtain ideal parameter estimates in terms of this "least squares" criterion. WLS estimation is similar, but involves a variable that serves as a weight of all of the observations used in the sum of squared errors. Each squared error is weighted by this additional variable. Observations with greater weights are assigned greater influence in the estimation process, while observations with lesser weights are assigned lesser influence in the estimation process. An ideal choice of weighting variable would be the inverse of the actual variation for each response. Used this way, responses with greater associated variability (and therefore less certainty) would be given less weight in the estimation. Responses with lesser associated variability (and therefore greater certainty) would be given greater weight in the estimation.

Using weights within least squares estimation allows researchers to control the magnitude of the influence of each response in terms of estimating effects and also standard errors. Such weights can allow the data analyst to assign lower influence to responses with higher associated variation, therefore correcting the problems in parameter estimates and also hypothesis tests resulting from nonhomogeneous response variation.

4.4.2 Selecting the Weights

In some rare cases it is possible to obtain estimates of the variation for each response, which can then be used as weights in WLS. For example, consider a study intended to find associations between classroom instructor experience and student test scores. To protect individual student privacy, the researchers aggregate student test scores into class averages, and use these averages as the responses in a regression model. The researchers can then use the class sample variance (or the variance divided by the square root of the class size) as an estimate for the variation in each response. The inverse of the class variances can then be used as weights within WLS estimation to accommodate any nonconstant variation in class average test scores.

In many cases, it is not possible to construct a reasonable estimate of individual response variation. In such cases WLS can still be applied using another variable from the data as a proxy for changing variation. This selection of the weighting variable is often made using residual plots, in which the residuals are plotted against each variable in the data set (other than the outcome of interest). The plot showing the strongest or clearest association between residual variation and values of the predictor indicates which predictor should be used as the weight. Often predictors that were not included in the original regression model are considered as candidates for weighting.

4.5 SSOCS Analysis: Annual Suspensions

We are interested in using the School Survey on Crime and Safety data to predict the annual number of suspensions for insubordination reported by a school, using the following predictors.

- Uniforms: the existence of a school uniform policy
- Metal detectors: the presence of metal detectors for students to pass through
- Tipline: the availability of a "tipline" to report issues
- Counseling: the availability of counseling for students
- Crime: the level of crime in the area surrounding the school (low, moderate, or high)
- Discipline training: whether teachers have training available on discipline policies
- Behavioral training: whether teachers have training available for positive behavioral interventions
- Insubordinates: the annual number of insubordinate students
- Limited English: the percent of students with limited English language proficiency

We are generally interested in whether the collection of listed predictors can effectively describe the expected number of suspensions, and also whether each predictor individually has an association with suspensions. We will first analyze the data using a normal linear regression model, but, expecting issues with nonconstant variance, we will explore various transformations of the outcome and also the use of weighted least squares with various predictors as weights.

4.5.1 Exploratory Data Analysis

We presented exploratory analyses in Chapter 1, and a few of the relevant findings are presented for discussion here. First consider the distribution of the outcome of interest, i.e., annual school suspensions for insubordination. The mean number of annual suspensions is 7.852, with a median of 0, a discrepancy that suggests right-skewness in the data. The sample variance is 4863.123, which shows the variance is much larger than the mean. Figure 4.1 shows both the histogram (in the left panel) and the box plot (in the right panel) for annual school suspensions, with the three values greater than 500 removed for ease of interpretation. While these three large values were removed to allow the plots to show details of the smaller

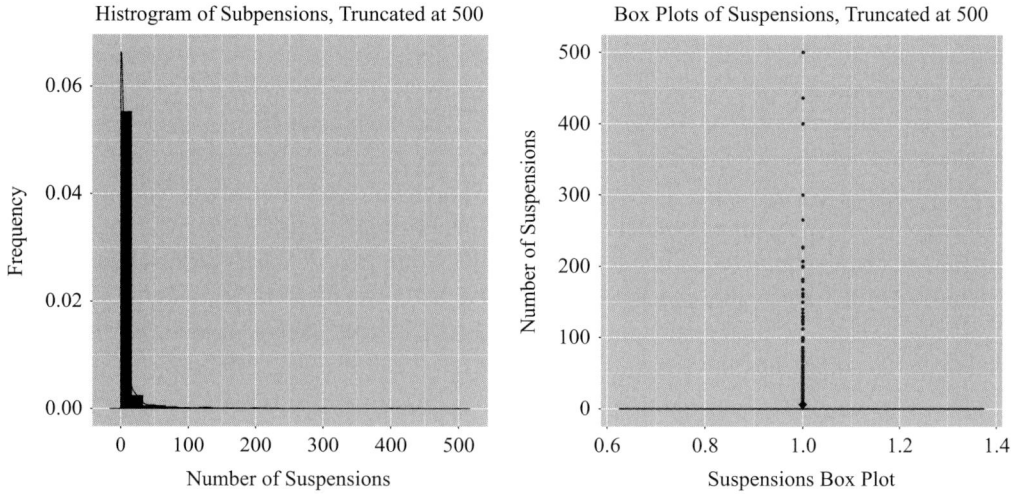

Figure 4.1 School Survey on Crime and Safety suspensions histogram (left panel) and box plot (right panel), with values greater than 500 suspensions omitted.

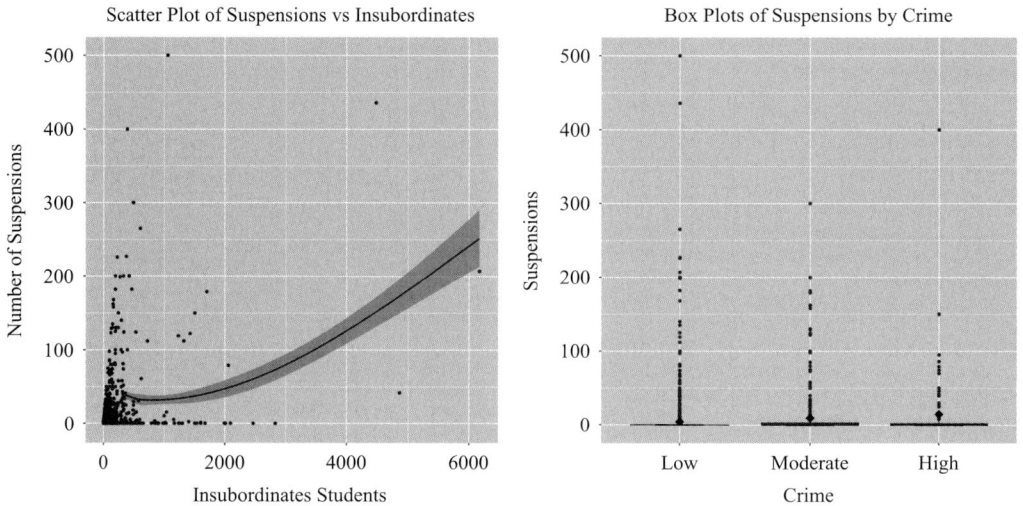

Figure 4.2 School Survey on Crime and Safety scatter plot of suspensions versus insubordinate students (left panel), box plot of suspensions by crime level (right panel), with values greater than 500 suspensions omitted.

values, all data are included in subsequent calculations and models. Both plots support the descriptive evidence of a right-skewed response with extremely large, positive values. This type of right-skewness is common for responses with values that must be 0 or larger.

Scatter plots of annual suspensions versus continuous predictors generally show similar patterns; there may be evidence of a relationship between suspensions and the plotted predictor, but most of the values are gathered near the origin. Similarly, box plots of annual suspensions grouped by categorical predictors show an outcome for each group that is clearly skewed to the right. A scatter plot of suspensions versus number of insubordinate students is shown in the left panel of Figure 4.2, omitting schools with more than 500 suspensions during the year of interest. Box plots of annual suspensions by crime level, for 500 or fewer annual suspensions, are shown in the right panel of Figure 4.2, also omitting schools with greater than 500 suspensions.

4.5.2 Normal Linear Model

We begin with a normal linear regression model. All the predictors are included along with multiplicative interactions of both discipline training and behavioral training with area crime level. Through such interactions we are able to determine whether the differences in expected annual suspensions between schools with and without the two types of training changes depending on area crime level. In other words, we are able to determine whether the effectiveness of the two types of training differs across area crime levels. Predictions of mean annual suspensions are made using the prediction function

$$
\begin{aligned}
\mu_{\text{suspensions}} = {}& \beta_0 + \beta_1(\text{uniforms}) + \beta_2(\text{metal detectors}) + \beta_3(\text{tipline}) \\
& + \beta_4(\text{counseling}) + \beta_5(\text{moderate crime}) + \beta_6(\text{high crime}) \\
& + \beta_7(\text{discipline training}) + \beta_8(\text{behavioral training}) \\
& + \beta_9(\text{insubordinates}) + \beta_{10}(\text{limited English}) \\
& + \beta_{11}(\text{discipline training} \times \text{moderate crime}) \\
& + \beta_{12}(\text{discipline training} \times \text{high crime}) \\
& + \beta_{13}(\text{behavioral training} \times \text{moderate crime}) \\
& + \beta_{14}(\text{behavioral training} \times \text{high crime}),
\end{aligned}
\tag{4.1}
$$

where $\mu_{\text{Suspensions}}$ represents the mean number of annual suspensions for an individual school. Within the prediction equation, we included indicators for moderate and high crime locations, implicitly using low crime as the reference for comparisons. As we presented in Chapter 3, the test for nonconstant variance in the normal linear model gives a test statistic of $X^2 = 199\,732.500$, with associated p-value significant at any level ($p < 0.001$). Thus, we have evidence that the variation in the number of suspensions is not constant across predictor values.

4.5.3 Outcome Transformations

Figure 4.3 shows a plot of residuals versus model fitted values, using the residuals from the normal linear model of annual suspensions. The straight-line "funnel" shape of the plot suggests either a square-root transformation or a logarithmic transformation may be effective in minimizing the effects of the nonconstant response variation. Figure 4.4 shows the resulting plots of residuals versus fitted values for both the model with a square root-transformed response, and the model with a natural-logarithm-transformed response. That is, the values of annual suspensions in the data are replaced with $\sqrt{\text{suspensions}}$ for the first transformation. For the second transformation, the values of annual suspensions are

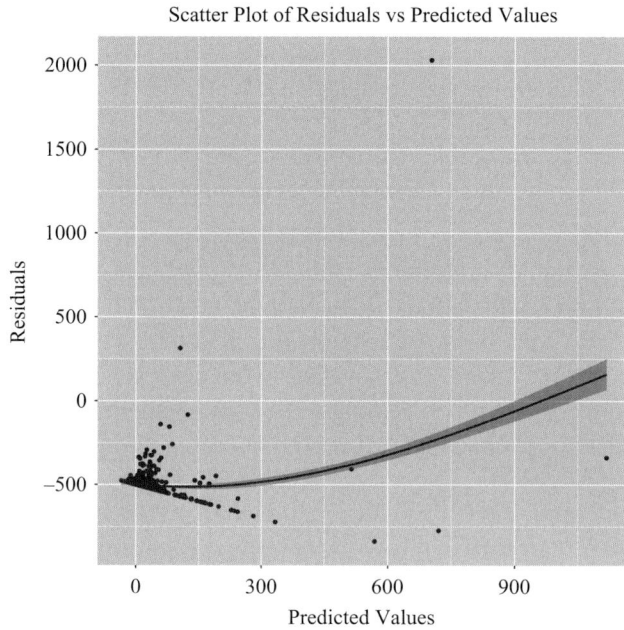

Figure 4.3 School Survey on Crime and Safety suspensions normal linear model residuals versus predicted values.

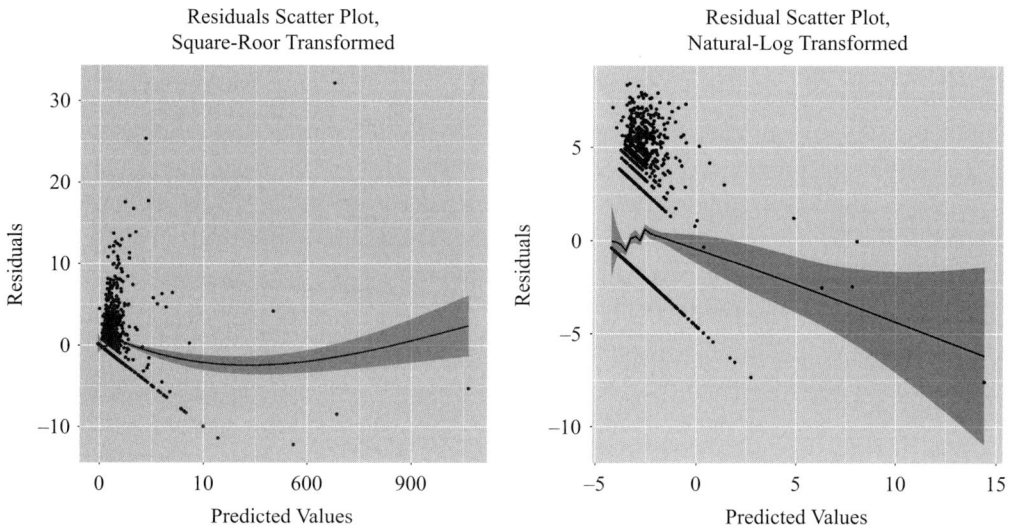

Figure 4.4 School Survey on Crime and Safety plot of residuals versus predicted values for square-root-transformed suspensions (left panel), and for natural-logarithm-transformed suspensions (right panel).

replaced with ln(suspensions + 0.01), where the adjustment by 0.01 is used to avoid taking the logarithm of zero, which is an undefined operation.

The residual plot for the square-root-transformed response model, shown in the left panel of Figure 4.4, depicts a better pattern than the associated plot for the original normal linear model, as there is clear vertical dispersion of residuals for small fitted values. However, the plot shows a diagonal "line" below which no residual values are observed. This line in the residual plot is common in data analysis situation in which the outcome of interest is truncated and right-skewed. Such a shape indicates that estimated values from a model are often underestimating comparable observed values (hence large residuals) but not overestimating as compared to observed values (hence no negative residuals at certain places in the plot). This is true for our current data analysis situation, as annual school suspensions are naturally truncated (there can be no "negative" numbers of suspensions) and the exploratory data analysis showed the outcome to be heavily right-skewed.

Similarly, the residual plot for the natural-logarithm-transformed response model, shown in the right panel of Figure 4.4, shows the dispersion in the residuals more clearly than the associated plot from the normal linear model, but again the boundary "line" in the residual plot is present, suggesting predicted values still tend to underestimate those observed. The corresponding tests of constant variance for the square root-transformed response model ($X^2 = 8\,285.980$, $p < 0.001$) and the natural-logarithm-transformed response model ($X^2 = 177.040$, $p < 0.001$) both provide evidence of heteroscedasticity in the model residuals. While both homoscedasticity tests are significant, the estimated χ^2 values are smaller than the value provided by the normal linear model, suggesting an improvement in the level of heteroscedasticity.

4.5.4 Weighted Least Squares

Because the method of transforming the response did not properly account for the heteroscedasticity in the data, we applied the weighted least squares approach to account for nonconstant variance. Based on the residual plots from the original normal regression model, shown in Figure 4.5, some predictors can be used to describe the changing variation in the normal model residuals. The left panel of the figure shows normal model residuals versus the number of insubordinate students, which shows the straight-line funnel shape seen in earlier residual plots. This means values of insubordination can give an indication of the amount of dispersion we would expect to see in the normal linear model residuals.

The right panel of Figure 4.5 shows it is possible that the percent of students below the 15th percentile in standardized testing can explain changes in residual variation, as the vertical fluctuation in residuals appears to diminish for larger values of this predictor. While the percent of students below the 15th percentile in standardized testing was not included in the original analyses, it can be used as a weight when applying weighted least squares.

Figure 4.6 shows the plots of residuals versus fitted values for both the model using the number of insubordinate students as weights, and the model using the percent below the 15th percentile in standardized testing as weights. Because the number of insubordinate students showed a positive relationship with residual variation (larger values of insubordinates showed greater fluctuation in residuals in Figure 4.5), the inverse

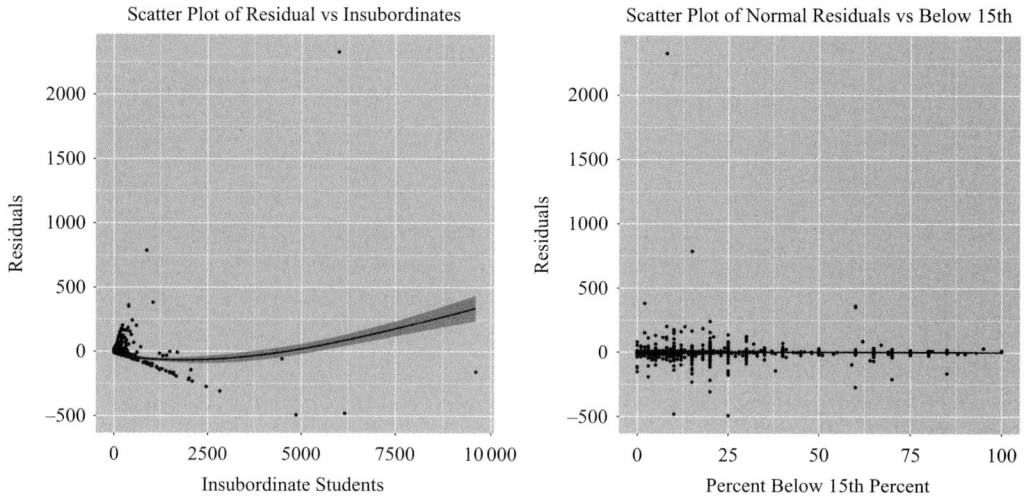

Figure 4.5 School Survey on Crime and Safety normal linear model for suspensions, residuals versus number of insubordinate students (left panel), residuals versus percent below 15th percentile on standardized tests (right panel).

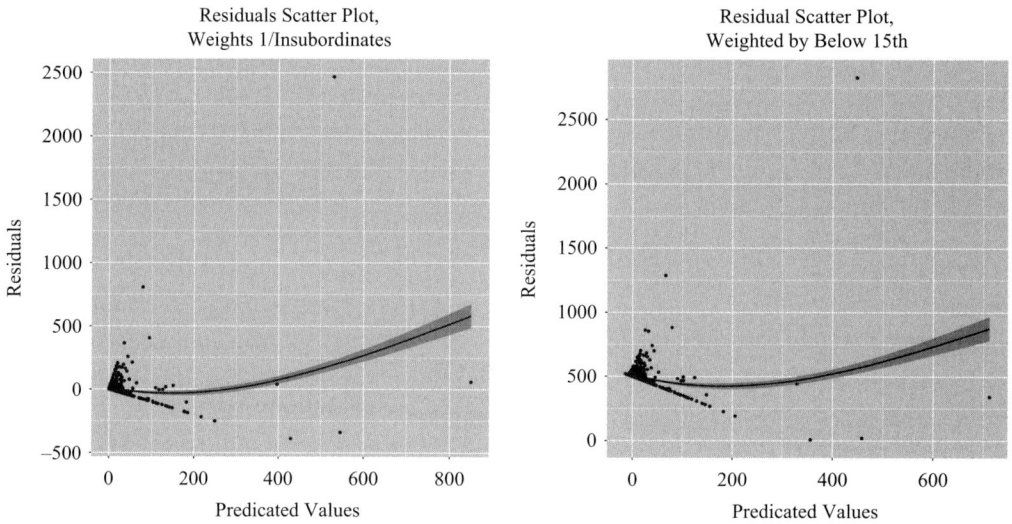

Figure 4.6 School Survey on Crime and Safety suspensions model residuals versus predicted values, weighted by the inverse of the number of insubordinate students (left panel), weighted by the percent of student below the 15th percentile on standardized tests (right panel).

$(1/(\text{insubordinates} + 0.01))$ was used as the modeling weight. We used the adjustment of 0.01 to avoid division by zero. The residual plots from both panels continue to show heteroscedasticity similar to the original normal linear model.

Results of the nonconstant variance test, presented in Table 4.1, show both the model using percent below the 15th percentile as weights ($X^2 = 124\,497.000$, $p < 0.001$) and the model using the inverse of insubordinates as weights ($X^2 = 11\,286.460$, $p < 0.001$) to be

Table 4.1 *School Survey on Crime and Safety evidence of heteroscedasticity in suspensions modeling options.*

Model	Homoscedasticity	
	Test statistic	*p*-value
Normal linear	199 732.500	< 0.001
Square root transformed	8285.980	< 0.001
Logarithm-transformed	177.040	< 0.001
WLS, percent below 15th Percentile	124 497.000	< 0.001
WLS, inverse insubordination	11 286.460	< 0.001
Logarithm-transformed WLS, inverse insubordination	4.052	0.044

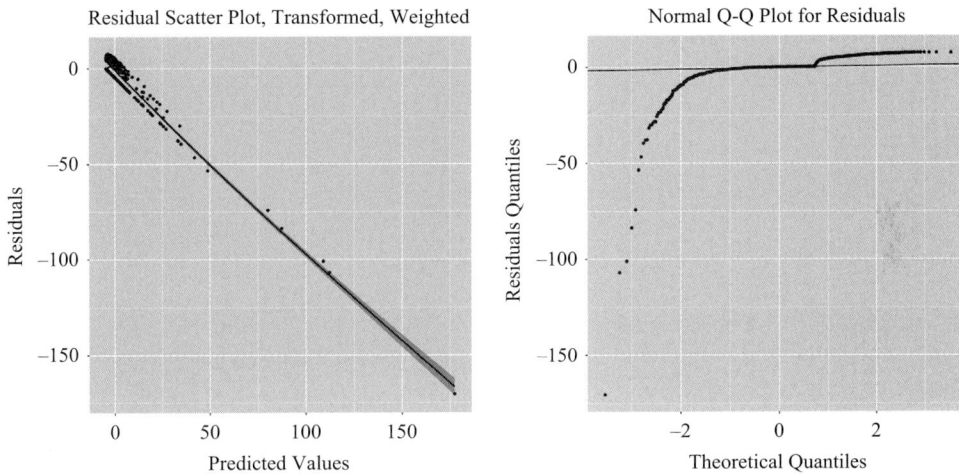

Figure 4.7 School Survey on Crime and Safety model for natural-logarithm-transformed suspensions weighted by inverse number of insubordinate students, scatter plot of residuals versus predicted values (left panel), normal Q-Q plot (right panel).

improvements on the normal linear model. However, neither weighted least squares model showed as much improvement as the model using a natural-logarithm-transformed response.

In order to improve the performance of the proposed transformed or weighted models with respect to the issue of nonconstant variance, we performed a final analysis combining the two approaches of outcome transformation of weighted least squares. Applying the inverse of insubordination as weights to a model with a natural-logarithm-transformed response, the test of homoscedasticity fails to show evidence of nonconstant variance ($X^2 = 4.052$, $p = 0.044$). The residual plot shown in the left panel of Figure 4.7 shows a clear pattern of residuals versus fitted values, suggesting the need for additional predictors to adequately capture the variation in annual suspensions. However, the plot does not show strong evidence of changes in variation. While the normal probability plot in the right panel of Figure 4.7 shows clear evidence of right-skewness in the data, we will use the

Table 4.2 *School Survey on Crime and Safety results of regression of natural-logarithm-transformed annual suspensions, weighted by inverse number of insubordinate students.*

Predictor	Estimate	Std. error	t-value	$Pr(>|t|)$
Intercept	−4.603	0.001	−449.178	< 0.001
Uniforms	−0.000003	0.008	< 0.001	> 0.999
Metal detectors	−0.002	0.020	−0.084	0.933
Tipline	0.001	0.006	0.192	0.848
Counseling	0.002	0.009	0.213	0.831
Moderate crime	−0.001	0.023	−0.053	0.958
High crime	−0.003	0.035	−0.092	0.927
Discipline training	−0.0002	0.007	−0.037	0.970
Behavioral training	−0.002	0.008	−0.199	0.842
Insubordinates	0.019	0.001	14.297	< 0.001
Percent limited English	−0.00003	0.0002	−0.166	0.868
Discipline × moderate crime	0.002	0.021	0.081	0.935
Discipline × high crime	0.001	0.026	0.058	0.954
Behavioral × moderate crime	0.004	0.024	0.170	0.865
Behavioral × high crime	0.004	0.035	0.118	0.906

MSE	0.398	
R^2	0.075	
Adjusted R^2	0.069	

natural-logarithm-transformed weighted least squares model to make interpretations. We present more advanced methods to handle such right-skewed data in Chapter 6.

4.5.5 Parameter Interpretations

Results of fitting the model are shown in Table 4.2. The one clearly significant predictor is the number of insubordinate students, with parameter estimate 0.019. Interpretations for model effects are challenging after applying response transformations. Very generally, the positive effect suggests the mean number of annual suspensions is expected to increase with an increase in insubordinate students, assuming all other predictors in the model are held constant. Specifically, the mean of the natural logarithm of number of suspensions is expected to increase by 0.019 for each additional insubordinate student. Recall that such an interpretation is a comparison between populations, and not a statement about the change of an individual school. The interpretation means that, when comparing two similar populations of schools with respect to all other predictors, the population of schools with an average number of insubordinate students that is greater than the other population is expected to have a higher average number of annual suspensions.

The nonsignificant parameters can be interpreted. Based on the model, the mean number annual suspensions is not expected to differ meaningfully between schools with and without uniform policies, metal detectors, or student counseling services.

Table 4.3 *School Survey on Crime and Safety predicted mean annual suspensions for schools with 0 to 500 insubordinate students.*

	Insubordinate students					
	0	50	100	150	200	250
Log (annual suspensions)	−4.599	−3.652	−2.706	−1.759	−0.812	0.134
Annual suspensions	0.010	0.026	0.067	0.172	0.444	1.143

	Insubordinate students				
	300	350	400	450	500
Log (annual suspensions)	1.081	2.072	2.974	3.920	4.867
Annual suspensions	2.947	7.593	19.568	50.425	129.940

4.5.6 Model Prediction

Mean values can be predicted from the model selected, but we need to remember that the outcome modeled was the logarithm-transformed annual suspensions. The values predicted by the model represent the mean of log(annual suspensions),

$$\hat{\mu}_{\text{log-susp}} = -4.603 - 0.000003(\text{uniforms}) + \cdots + 0.0004(\text{behavioral training} \times \text{high crime}). \tag{4.2}$$

The predicted values for log(annual suspensions) can be anti-logged to return to the original scale of annual suspensions, but the anti-log of a mean of log-annual suspensions does not necessarily produce the desired mean of annual suspensions. We present ways to predict means of count data more appropriately in Chapter 6. Predictions based on the current model for various numbers of insubordinate students are shown in Table 4.3, assuming schools have a uniform policy, no metal detectors, no tipline, counseling, moderate crime in the location of the school, discipline and behavioral training for teachers, and an average percent of students with limited English proficiency. Notice the model does not predict more than 1 annual suspension until the number of insubordinate students exceeds around 250, but increases quickly beyond that point. The relationship in the model is nonlinear, and therefore the predicted values change at a rate faster than could be seen using a straight-line relationship. This is apparent from the plot of predicted values overlayed on observed values, shown in Figure 4.8. This plot shows that, while predicted numbers of suspensions do increase with increased numbers of insubordinate students, most predictions suffer from under-estimation of actual counts of suspensions.

Overall we have found the normal linear model to be inappropriate when using the number of suspensions as the response, specifically in violating the assumption of constant variance. Neither applying common transformations of the response nor using other variables as weights were able to correct the violations of the constant variance assumption alone. The combination of applying a transformation *and* applying weights in the estimation produced a model that satisfies the homogeneity of variance assumption. The number of suspensions was shown to increase with the number of insubordinate students, but the interpretation remains unclear because of the transformation applied.

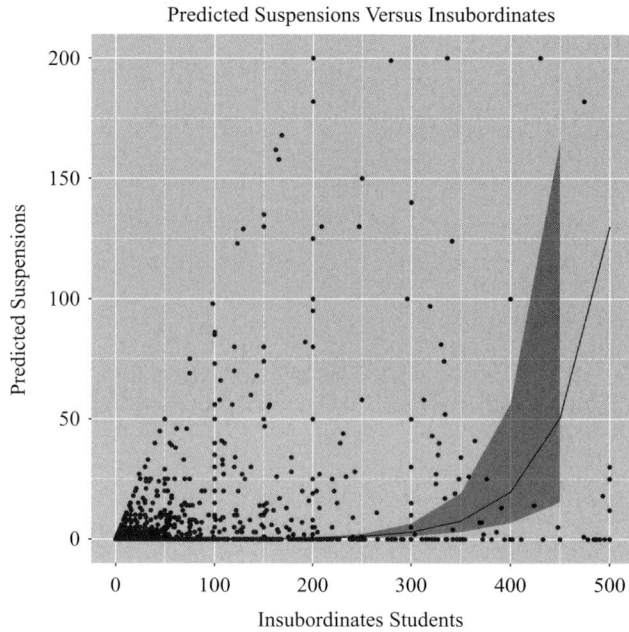

Figure 4.8 School Survey on Crime and Safety plot of suspensions versus insubordinate students, with smooth predicted curve.

4.6 Fire-Climate Analysis: Decade Averages

Using the Fire-Climate Interactions data, we are interested in modeling the decade averages of fire indicators using as predictors the following variables.

- Decade: the decade of observation
- Region: the region of observation (Pacific Northwest, Interior West, and Southwest)

Each decade average is calculated as the average number of fire indications across ten years, with a single calculated average per decade for each site within each region. Therefore the data have been aggregated by collapsing many observations into one summary value, the average. In many cases such aggregated data can show heterogeneous variation due to the different numbers of raw observations incorporated into each average. When this occurs, other aggregated values, such as the decade variation, are often turned to as weights in the analysis. We will proceed by first analyzing the data using a normal linear regression model. Then we will turn to the use of weighted least squares with another summary measure, the variation in fire indications, used as weights in the estimation.

4.6.1 Exploratory Data Analysis

Exploratory analyses were presented in Chapter 1. Some relevant descriptives are investigated here to guide the modeling approach. The five-number summary of decade averages shows a minimum average of 1 fire event and a maximum average of 4.7 fire events. The median remains at 1, while the mean is 1.37, providing some evidence of

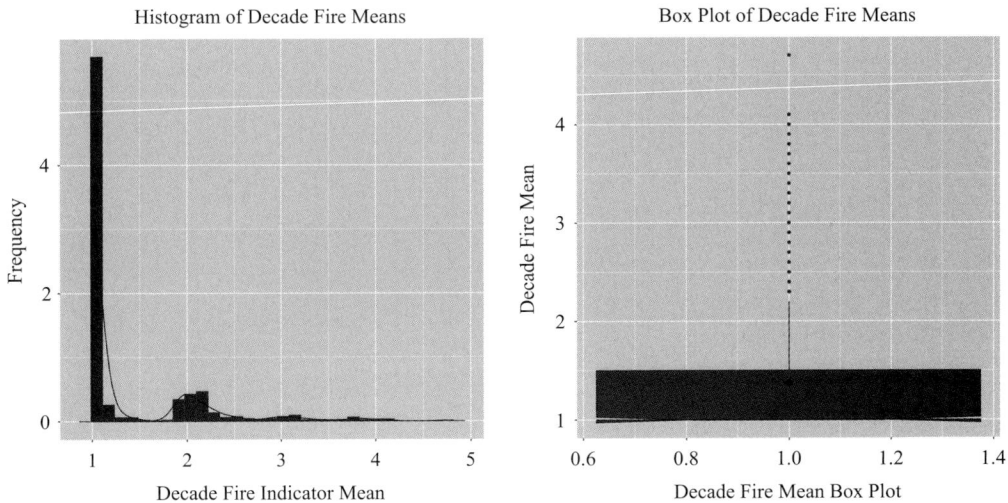

Figure 4.9 Fire-Climate Interactions fire count decade averages histogram (left panel) and box plot (right panel).

right-skewness in the response. The variance of decade averages is 0.438. The left panel in Figure 4.9 shows a histogram of decade averages, which shows clear evidence of skewness to the right in the data. The box plot shown in the right panel of Figure 4.9 supports the same conclusion.

Turning next to the relationship between decade averages and predictors, the left frame of Figure 4.10 shows a scatter plot of decade averages plotted versus decade. The data do not suggest a strong linear trend of decade averages over time, but possibly a quadratic trend as values appear to descend on either side of a peak at around 1750. This scatter plot also shows that the vertical dispersion in decade averages appears to change across decades, suggesting nonconstant variation. The second frame of Figure 4.10 shows box plots of decade averages for each region. The average outcome for the Pacific Northwest region appears to be greater than that of both other regions, and the variation in the response appears to be greatest for the Pacific Northwest region. The Interior West region shows the smallest variation. These box plots support existing evidence that the variation in decade averages is not consistent across predictor values.

4.6.2 Normal Linear Model

We begin with a normal linear regression model, including indicators for two of the regions to serve as comparisons to the reference region (the Interior West region), along with both linear and quadratic terms for decade. Predictions of the typical decade average are then made using the following prediction function:

$$\mu_D = \beta_0 + \beta_1(\text{PNW}) + \beta_2(\text{SW}) + \beta_3(\text{decade}) + \beta_4(\text{decade}^2), \qquad (4.3)$$

where μ_D indicates the mean of decade averages, and "PNW" and "SW" indicate Pacific Northwest and Southwest regions, respectively. As we discussed in Chapter 3, the statistic for the test of homogeneity of variance for this normal linear regression model is 208.00

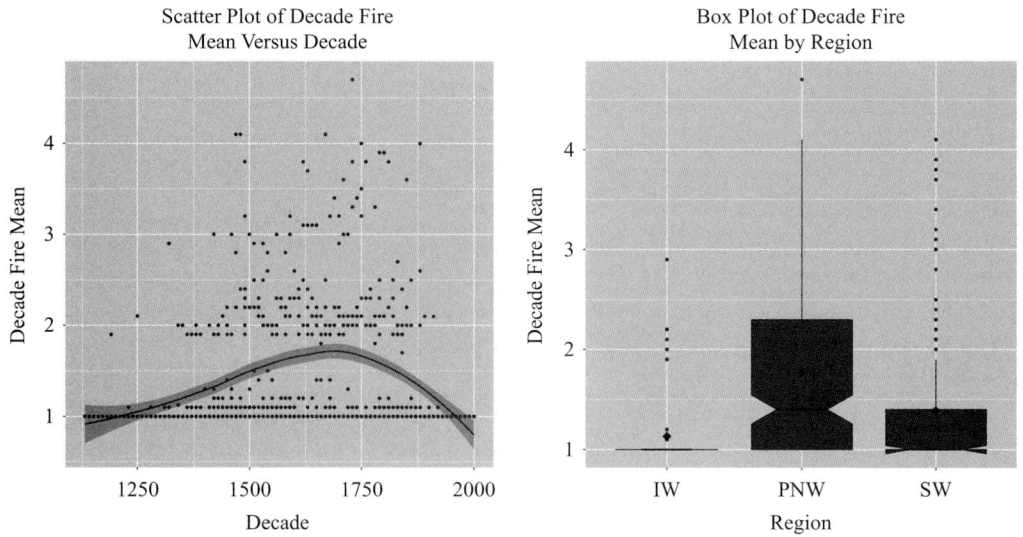

Figure 4.10 Fire-Climate Interactions scatter plot of decade average fire counts versus decade (left panel), box plots of decade average fire counts by region (right panel).

with associated p-value $p < 0.001$. The test shows clear evidence of heterogeneous variation in the outcome. The plots in Figure 4.11 show the plots of residuals from the normal model versus predicted values and also versus decade values. The plot in the left frame, of residuals versus predicted values, clearly shows a funnel shape in which the variation in the residuals increases with predicted values. This residual plot also shows the lower boundary diagonal "line" below which no residuals are observed, suggesting that predicted values are failing to adequately predict unusually large observed values. The scatter plot of residuals versus decade in the right frame of Figure 4.11 shows similar patterns, with curvature reflecting the quadratic term in the model.

4.6.3 Weighted Least Squares

Weighted least squares is commonly applied to outcomes that have been calculated as some kind of aggregated value, such as the decade averages used in our current analysis. In this case the decade variances have also been calculated, and provide a natural weight to be used as a representation of the variation in the outcome of interest. Figure 4.12 shows a residual scatter plot using residuals from a regression model fit using $1/(\text{decade variance} + 0.01)^2$ as weights in the estimation. (The 0.01 term is added to avoid dividing by 0.) The residual plot versus predicted values shows an improvement over the normal linear regression model, but still shows evidence of increasing variation as predicted values increase. The statistic for the test of homogeneity of variance using the weighted model is 110.96 with p-value $p < 0.001$, also suggesting violations of constant variance.

While the test statistic and residual scatter plot both show improvement, the heterogeneous variance problem has not been completely addressed using weighted least squares. The performance of the model may be improved by using higher powers of the decade

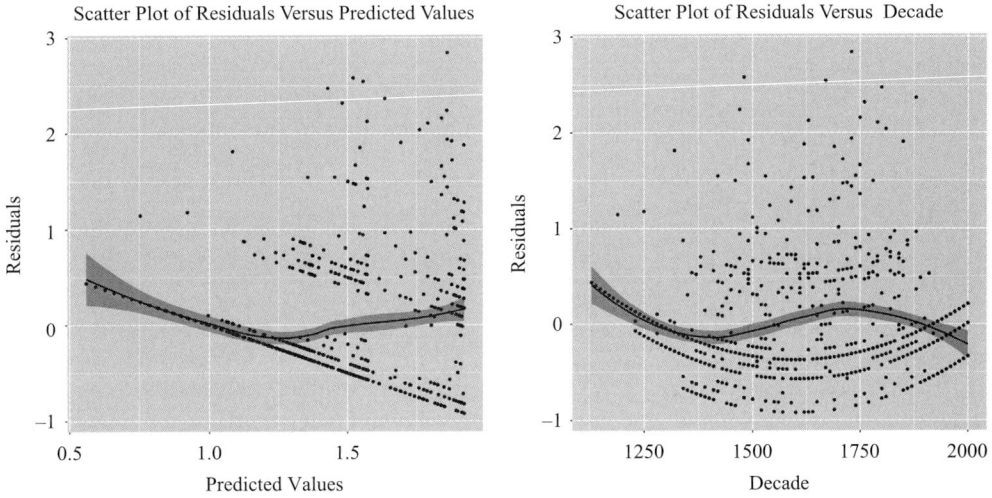

Figure 4.11 Fire-Climate Interactions scatter plot of normal linear model residuals versus predicted values (left panel), scatter plot of normal linear model residuals versus decade (right panel).

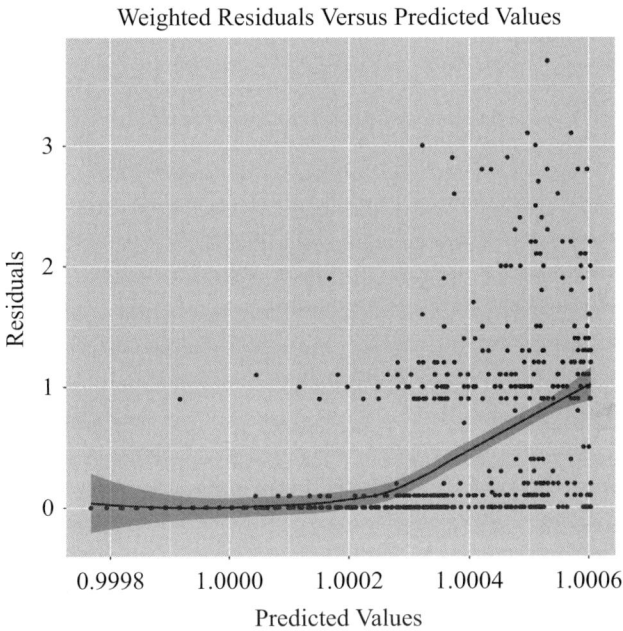

Figure 4.12 Fire-Climate Interactions decade average fire count model, weighted by decade variation in fire counts, residuals versus predicted values.

variance as weights, or by applying transformations to the response such as the logarithmic transformation. However, additional transformations lead to challenging interpretations of the estimated model coefficients. A more appropriate analysis of the Fire-Climate Interactions data is provided in Chapter 6.

Table 4.4 *Fire-Climate Interaction results of regression of decade fire averages, weighted by decade fire variances.*

Predictor	Estimate	Std. error	t-value	$Pr(>|t|)$
Intercept	9.929	0.008	120.908	< 0.001
Pacific Northwest Region	0.0002	0.0006	0.430	0.667
Southwest Region	0.0002	0.0004	0.368	0.713
Decade	0.00001	0.00001	0.899	0.369
Decade^2	$-3\text{e-}9$	< 0.001	-0.910	0.363

	MSE	0.159
	R^2	0.001
	Adjusted R^2	-0.004

Table 4.5 *Fire-Climate Interaction predicted decade average fire indications, by region.*

Model	Region		
	Pacific Northwest	Interior West	Southwest
Weighted model	1.0006	1.0003	1.0005
Normal model	1.9213	1.3738	1.5740

4.6.4 Parameter Interpretations

Results of fitting the model using weighted least squares are provided in Table 4.4. None of the predictors show significance in the model. The data fail to provide evidence of a difference in expected decade average fire indicators between Interior West, Pacific Northwest, and Southwest regions, accounting for quadratic differences across decades. The data also do not provide evidence of changes in decade averages across decades. The value $R^2 \approx 0.001$ indicates that less than 1% of the variation in decade averages is accounted for by the weighted least squares linear model, which is extremely low explanatory ability.

4.6.5 Model Prediction

In spite of nonsignificant results, the model can be used to predict decade averages of fire indicators for different regions. Table 4.5 shows the predicted decade fire averages for the three regions, evaluated at the mean decade value (1 606.684), for both the normal linear regression model and the model weighted by decade variance. Predictions using the weighted model show very little difference across regions, while the predictions from the normal regression model more closely reflect the observations made using the box plots. Superior predictions of fire indicators are made in Chapter 6 and also in Chapter 8.

Overall the model for mean fire counts using the variation in fire counts as weights showed improvement in the violations of the constant variance assumption, but the assumption was not completely addressed through such weights. The model failed to show evidence of differences across the three regions under consideration, and produced predictions that may be less informative than raw averages.

4.7 Summary

In this chapter we examined the use of transformations and weighted estimation approaches to address the issue of heterogeneous variation in the outcome of interest. In some cases the application of transformations, weighted estimation, or some combination of the two can alleviate model issues with constant variation, but interpretations of transformed responses are challenging and selection of both transformations and weights remains arbitrary. In general we believe the methods presented in subsequent chapters to be more appropriate for addressing the types of responses that often show nonconstant variation.

4.8 Further Reading

A classical treatment of linear models with transformations and weighted estimation can be found in Kutner et al. (2004). Some nice texts explaining the use and application of the linear model are Freedman (2009), Glass and Hopkins (1995), and Agresti and Finlay (2008).

5

Discrete, Categorical Response Models

5.1 Categorical Responses

Data collection often involves classifying individual observations into mutually exclusive categories. For example, industrial experiments can involve classifications of product testing as a success or failure; education studies can classify students as unsatisfactory, partially proficient, proficient, or advanced; health studies often identify diseases as present or not present. When such types of information are considered as responses in a statistical model, normal regression techniques are insufficient as categorical responses cannot be assumed to behave normally, as we saw with the poor fit of some of the normal regression models from Chapter 3. In fact, categorical data routinely involve highly skewed data, nonconstant variation in the response, and nonlinear relationships with predictors. As a consequence, researchers prefer to use nonlinear logistic regression modeling in lieu of normal least squares methods.

When modeling categorical responses it is most common to model the probability that an observational unit is classified according to any of the possible response groups. With two possible classifications, a binary response, ordinary logistic regression is most often applied to model the probability associated with one of the two outcome groups. Alternative methods exist, but have disadvantages in interpretations as compared to the logistic regression model. When considering a nominal response with more than two response classifications, multinomial logistic regression can be applied. When considering an ordinal response with classifications that can be arranged by some measure of magnitude, a number of ordinal multinomial logistic regression models can be used.

While all logistic models involve probabilities associated with different response classifications, interpretations and significance are often made with respect to probability comparison statistics such as odds and odds ratios. Logistic regression models allow for nonlinear relationships between response probabilities and predictors, and allow for nonconstant variation in the response, although the nonconstant variation is expected to behave according to a specific relationship with the response probability.

5.2 Binary Logistic Regression

Consider the case of a binary response variable, such as pass versus fail, approved versus denied, or operational versus damaged. Binary logistic regression can be used to model the probability associated with one of the response classifications, often referred to as the "success" classification. (The choice is arbitrary and does not affect analysis; the probability associated with the other group will simply be the complement of the "success" probability modeled. Model fit, effect size, and variable significance all remain identical.) Logistic regression allows for nonnormally distributed responses, for nonlinear relationships with predictors, for nonconstant response variation, and can accommodate continuous, binary, and categorical predictors. It is important to keep in mind the nature of the response: instead of modeling a traditional mean as in normal regression, logistic regression models the probability of a specific outcome.

5.2.1 Descriptive Statistics for Binary Outcomes

Some types of exploratory analyses are preferred over others for categorical data. For binary responses, scatter plots and means remain useful. In addition, analysts often investigate relevant proportions through contingency tables, bar plots, and the logistic histogram scatter plot.

When exploring the relationship between a binary outcome and a continuous independent variable, a scatter plot of the response versus the predictor can be constructed, but interpretation of the plot is challenging. Binary outcomes are typically represented numerically as 0 or 1 values, and consequently the scatter plot shows all responses along the horizontal line, either at 0 or at 1 on the vertical, making it difficult to determine the strength of any relationship. Instead it is common to use a logistic histogram plot, in which histograms, or shaded blocks indicating magnitude, associated with both the 0 and 1 outcomes are plotted on the same graph at 0 and at 1 on the vertical, respectively. Such a plot can be used to identify values of the predictor on the horizontal for which there are greater or fewer 0 or 1 outcomes.

Plots can be very helpful in identifying *separation* in the binary response. Separation (or complete separation) refers to the phenomenon of a binary response classified perfectly according to a predictor. For example, consider the case of modeling the probability of high-school students passing a standardized exam. If all students above a certain GPA pass the standardized exam, while all students below that GPA fail the exam, then GPA perfectly classifies passing the results of the exam. There is no need for a statistical model, and in fact fitting a logistic model will suffer from convergence failure. To avoid such a problem, the analyst can look to the logistic histogram plot to make sure the 0 and 1 values have some overlap with respect to all continuous predictors. More commonly, it is possible to have quasi-complete separation. This refers to the situation in which there is some overlap in 0 and 1 outcomes, but the overlap is very limited, which causes values estimated from a logistic regression model to become unreliable.

When exploring the relationship between a binary outcome and a categorical predictor, contingency tables and bar plots prove more useful than scatter plots. Typically, a contingency table is constructed, with a column for each possible response and a row for each possible predictor classification, and counts of instances of each are filled in. Structural zeros can be seen by 0 counts in any of the table cells, suggesting that no instances of the associated outcome category occur for that predictor category. Separation can be identified by positive counts within one or a few of the predictor categories, and zeros in other columns of the table. To avoid such structural zero or separation problems, the analyst looks for positive counts in all cells of the contingency table.

Data analysts occasionally produce bar plots in which a categorical predictor's levels are listed along the horizontal, and the count or proportion of successes is represented by the height of the bar. Colors or shading in the bars can be used to indicate counts or proportions of successes according to another, third variable. In this way analysts can form a quick, descriptive opinion of whether the probability of success varies across levels of a predictor.

5.2.2 The Logistic Regression Model

The basic notation of logistic regression models was introduced in Chapter 2. To maintain meaningful predictions of response probabilities, which must produce values between 0 and 1, logistic regression imposes a nonlinear relationship between the response probability and the predictors,

$$\pi_i = \frac{\exp\left(\beta_0 + \sum_{j=1}^{J} \beta_j x_{ij}\right)}{1 + \exp\left(\beta_0 + \sum_{j=1}^{J} \beta_j x_{ij}\right)}, \tag{5.1}$$

where π_i is the probability of success for observational unit i, exp() represents the natural exponentiatial function (or the anti-log function), and the β_j and x_{ij} represent regression parameters and predictor values as with normal regression. In this expression the x_{ij} are observed data, while the β_j are unknown effects that must be estimated to produce values of π_i. The logistic regression model is often written equivalently in terms of a transformation of the response probability so that the right-hand side of the expression, involving the predictors, reminds of normal linear regression,

$$\ln\left(\frac{\pi_i}{1 - \pi_i}\right) = \beta_0 + \sum_{j=1}^{J} \beta_j x_{ij}, \tag{5.2}$$

where ln() represents the natural logarithm function. Notice the right-hand side of this equation looks identical to that of a traditional normal linear model. The nonlinear expression on the left of the logistic regression equation is referred to as the "logit function" or the "log-odds" and represents a crucial component of interpretation and understanding logistic regression. Both expressions of the logistic regression model from Eq. (5.1) and Eq. (5.2) are equivalent. In both cases the x_{ij} represent observed predictor values, while the π_i and β_j are parameters to be estimated.

5.2.3 Interpreting Model Coefficients

When analyzing categorical data, analysts often calculate the "odds" associated with two categories. Written as a ratio, the odds represents the ratio of probabilities associated with two response categories. A calculated odds can be interpreted based on its value relative to a value of 1: an odds greater than 1 suggests the numerator event is more likely, while an odds less than 1 suggests the denominator event is more likely. For binary responses there is a single odds: the ratio of the probability of success to the ratio of the probability of failure. As a ratio of probabilities, odds will always be strictly nonnegative and right-skewed. Therefore odds can be expected to relate to arbitrary types of predictors through a log transformation. In this way the log-odds used in logistic regression is an intuitive choice to equate to the standard regression combination of parameters and predictors.

Because of the nonlinear nature of the relationship between the response probability and the regression parameters, interpretations based on a logistic regression model are made differently than with a normal linear regression model. Instead of interpreting a single regression parameter β_j as the expected difference in the mean response between populations, with logistic regression β_j has an interpretation as the expected difference in the log-odds between populations, which is not easy to explain. Typically, the regression parameter is anti-logged (or the natural exponential function is applied), $\exp(\beta_j)$, to give what is called the odds ratio. The odds ratio provides the expected *multiplicative* difference in the odds of success between two populations, that is, the multiple by which the original odds of success is expected to change as the predictor changes across populations. Wald tests of hypotheses provide evaluations of the significance of these effects.

Alternatives to logistic regression exist for modeling the probability of success for binary data. The linear probability model simply treats the probability of success as the left-hand side of a normal regression model. This can lead to predicted probabilities outside the acceptable range of 0 to 1. The probit regression model uses the inverse of the normal distribution function in place of the log-odds transformation. While the probit regression model has some benefits in terms of closed-form estimators, it eliminates the easy interpretations of the odds and odds ratios. The complementary log-log transformation is occasionally used in place of the log-odds, but again this eliminates the easy odds ratio interpretations. We suggest use of the logistic regression model in general.

5.2.4 Model Fit

Because of the nonlinear nature of the logistic regression model, OLS is no longer an appropriate method for estimation of parameters. Instead the iterative re-weighted least squares (IRLS) method is typically applied, which is an approximation to the method of maximum likelihood. The method of maximum likelihood is sensitive to issues such as separation and structural zeros, which is why these problems must be investigated through exploratory analyses. Before interpreting parameter estimates or performing hypothesis tests, it is important to assess the overall quality of the fit of any logistic regression model.

Once parameter estimates have been obtained it is necessary to determine whether the model fits the data well. Two common goodness-of-fit statistics are the Pearson χ^2 and the deviance χ^2. Both statistics are appropriate only in situations in which the predictors of the logistic regression model represent classifications (i.e., no continuous predictors can be present), and under certain regularity conditions are distributed as χ^2 variables with degrees of freedom equal to the difference between the number of predictor classifications and the number of model parameters. Significant test statistics suggest poor fit. In many situations, these statistics are reported descriptively as the value of the statistic divided by the degrees of freedom, which should be close to 1 for a model with good fit.

In addition to χ^2 statistics, a number of extensions of familiar linear model fit statistics have been constructed. A number of "pseudo R^2" statistics have been proposed to provide

Discrete, Categorical Response Models

Table 5.1   *Response classifications according to observation (columns) and model prediction (rows).*

		Observed outcomes	
		Failure	Success
Predicted outcomes	Failure	Specificity	
	Success		Sensitivity

summary descriptive statistics that remind researchers of those applied for linear regression models. The following are referred to as "pseudo" R^2 statistics because, while they are restricted to the range between 0 and 1 just as with traditional R^2, they are not directly tied to the estimation method as is true of linear regression models. Efron's pseudo R^2 is constructed similarly to the original R^2, in that it is calculated using a ratio of squared differences between observed and predicted values to the total variation in the response. Other more modern approaches use ratios of log-likelihood functions (part of the maximum likelihood process) with and without all of the predictors in the model.

In all cases, the value of any pseudo R^2 will be loosely interpreted as usual, as an approximate proportion of available information explained by the model. While pseudo R^2 cannot be taken to represent the proportion of response variation that is explained by the model, values closer to 1 are taken to suggest stronger fit. However, values of pseudo R^2 are commonly very small for logistic regression models. Because of the expected minimal magnitude and the unclear interpretation, we prefer to use statistics other than pseudo R^2 to assess fit of logistic regression models.

The most common method used to evaluate the goodness-of-fit for logistic regression models is the Hosmer-Lemeshow test. In applying the Hosmer-Lemeshow test, observations are partitioned into groups according to predicted success probabilities. For each group, the sum of predicted successes (using sums of predicted success probabilities) is compared to the sum of observed successes for that group, using a χ^2 goodness-of-fit test. A large statistic suggests that the predicted successes for all groups are not consistent with those observed, and therefore implies that the model does not fit the data well. While the Hosmer-Lemeshow test is almost always used to evaluate fit for a logistic regression model, its effectiveness is limited to data sets of about 25 000 observations or less; beyond this limit the test is fundamentally overpowered. The Hosmer-Lemeshow test suffers from subjectivity of its process, as the number of groups used is determined by the data analyst.

Another common method for evaluating the quality of fit for a logistic regression model relies on the model's ability to predict observed successes and failures. Often analysts will construct a response classification table, which is a contingency table in which the columns correspond to observed success or failure and the rows correspond to predicted success or failure, as shown in Table 5.1.

The upper-left entry in the table is the count of observed failures that are also classified as failures by the model. The associated probability of classifying an observed failure as a failure using the model is known as the specificity. The lower-right entry in the table is the count of observed successes that are also classified as successes by the model. The associated probability of classifying an observed success as a success using the model is known as the sensitivity. Obviously it is beneficial to have both high specificity and high sensitivity, while obtaining lower counts for the other two cells in the contingency table (representing errors in classification). Unfortunately, these classification tables require an arbitrary decision by the analyst. Because the logistic regression model produces predicted probabilities of success, in order to construct a response classification table it is necessary to select a cutoff value above which a predicted probability is considered to predict a success. While most software uses the cutoff value of 0.5 by default, the choice is arbitrary and can lead to misleading results, in particular for highly unusual events. The *receiver operating characteristic* considers all possible cutoff values.

The receiver operating characteristic (ROC) is constructed by calculating specificity and sensitivity values for all possible cutoffs. Typically, the results of these calculations are presented in an ROC curve, in which sensitivity is plotted against 1-specificity. In an ideal situation, as cutoffs increase, the sensitivity will increase without much of a decrease in 1-specificity. If the plot shows a diagonal line, it suggests that any increase in sensitivity is accompanied by a comparable loss in specificity. This is akin to making random guesses for the predictions of success or failure, and indicates a poorly fitting model. On the other hand, if the plot shows a dramatic increase in sensitivity without much loss in specificity, the shape will stretch far above the diagonal line in a concave-down shape. This suggests a model with very strong predictive ability, and correspondingly suggests good fit. In addition to the plot, researchers typically report the area under the ROC curve, which will be between 0.5 and 1. Values closer to 1 suggest stronger predictive power of the model. In practice, it is rare to see values much larger than 0.85 or 0.90, as classification this strong is often linked to separation problems and, consequently, with model estimation convergence failure.

In addition to evaluating the fit of a model using predictions of success and failure, numerous types of residuals have been proposed to assess fit. Unfortunately, raw and standardized residuals are often not used as extensively as with normal linear regression models. Typically, standardized deviance residuals are used to investigate observations that fit poorly with the model. Under certain assumptions, the deviance residuals will be approximately normally distributed. Therefore plots and descriptive statistics can be applied to determine whether there are unusually large deviance residuals, such as larger than 3 or 4 in magnitude. However, plots of deviance residuals often show patterns in logistic models, regardless of the quality of fit.

5.3 Nominal Multinomial Models

In some cases the response of interest may include more than two competing categories. For example, health researchers may be interested in the likelihood of patients being diagnosed with four or five potential illnesses; higher education researchers may be interested in the chance of different universities falling into different Carnegie Classifications; environmental researchers might be interested in predicting the probabilities that different nesting areas are associated with different species. In these cases the outcome of interest represents multiple mutually exclusive categories, and the interest remains in predicting the probability associated with each sampling unit being associated with each possible outcome category.

In such cases the binomial logistic regression model can be extended to one of many multinomial logistic regression models.

For cases in which the responses are nominal and *cannot* be inherently ordered, the most common modeling choice is the nominal multinomial logistic regression model. To construct such a model, a baseline or comparison outcome category is selected for comparison, and ordinary logistic regression models are constructed to compare each of the remaining outcome categories to this baseline category, using odds ratios. For example, in the case of modeling the probability that a patient is diagnosed as having good health, as having had myocardial infarction, or as having severe acid reflux, the researcher may choose "healthy" as the baseline category and construct two ordinary logistic regression models to predict both the odds of myocardial infarction versus healthy and the odds of severe acid reflux versus healthy. In general the model can be written

$$\ln\left(\frac{\pi_{ij}}{\pi_{iB}}\right) = \beta_{0j} + \sum_{k=1}^{K} \beta_{kj}x_{ik} \quad (j \neq B), \tag{5.3}$$

where π_{iB} is the probability that individual i is associated with the baseline category, π_{ij} is the associated probability for category j, x_{ik} represents the value of predictor k for individual i, while the coefficients β_{kj} depend on both the predictor x_k and also the outcome category of interest j. In this expression, the x_{ik} are observed predictor values while the β_{kj} are unknown effects, which must be estimated to produce values of each probability π_{ij}. Notice this model comprises multiple ordinary logistic models; for J outcome categories there will be $J-1$ ordinary logistic regression models. Every outcome other than the baseline category will have an associated logistic model.

Also notice that each individual will have an associated probability for each possible outcome category. For the example situation with three possible outcomes, each individual will have three associated predicted probabilities. Similar to the ordinary logistic regression model, these predicted probabilities can be written in terms of a nonlinear formula using the model predictors and parameters.

Coefficients can be interpreted similarly to those of ordinary logistic regression. The anti-logged coefficient, $\exp(\beta_{kj})$, represents the expected multiplicative change in the odds of outcome j versus the baseline, for a unit increase in predictor x_k. For nominal multinomial models it is very important to specify which outcome categories are being compared.

5.4 Ordinal Multinomial Models

For some research situations with categorical outcomes, the response categories have an inherent order. For example, education researchers may be interested in modeling the likelihood of elementary students performing as advanced, proficient, partially proficient, or unsatisfactory on standardized exams. Because each successive rating is considered to be "less than" the previous rating, the outcomes are ordinal. As another example, consider the interest of predicting different borrowing institutions' Moody's Long Term Credit Rating, from C (in default) through Aaa (prime). Because credit ratings can be ranked in increasing order of quality, the outcome of interest is considered to be ordinal. Using models that take advantage of the ordinal nature of the outcome of interest can produce statistical tests that are more powerful than traditional nominal models and provide answers to directional research interests.

There are three widely used models for predicting probabilities with ordered categorical outcome variables: the cumulative logit model, the adjacent categories model, and the continuation ratio model. Each model involves considering the ordered categories in a way that creates two groups to compare using ordinary logistic regression approaches. Unlike the nominal multinomial model, for ordinal outcomes it is common to apply a "proportional odds" model, meaning it is assumed that odds ratios are equivalent across outcome categories, regardless of which outcome categories are being compared. This means that the regression coefficients do not change depending on the response categories (thus omitting the j subscripts from the β_{kj} in the nominal multinomial model). The coefficients produced can be thought of as "averaged" over all possible comparisons of outcome categories. It is possible to fit any ordinal multinomial model without the proportional-odds assumption, but doing so runs the risk of failed convergence of parameter estimation, or predicted probabilities outside the acceptable range of $(0, 1)$.

5.4.1 Cumulative Logit Model

The cumulative logit model is constructed by calculating odds ratios comparing all categories greater than a given outcome group to all categories less than or equal to that outcome group. The model is similar to fitting all possible binary logistic regression models formed by collapsing response categories to form a binary outcome. The model can be written

$$\ln\left(\frac{\sum_{l=1}^{j} \pi_{il}}{\sum_{l=j+1}^{J} \pi_{il}}\right) = \beta_{0j} + \sum_{k=1}^{K} \beta_k x_{ki} \quad (j = 1, \ldots, J-1), \tag{5.4}$$

where the numerator within the logarithm represents all of the ordered response categories up to category j for individual i, the denominator represents all of the ordered responses categories greater than category j for individual i, and the coefficients β_k do not depend on the response category j (except for the intercept β_{0j}). Here the x_{ki} represent observed predictor values while the β_k are unknown effects that must be estimated through the model to predict values for the π_{ij}. For J total outcome categories, there will be $J-1$ equations, corresponding to the $J-1$ possible ways to split the ordered outcomes into two groups. Because of the proportional-odds assumption, the coefficients β_k can be interpreted very generally. The natural exponentiated (or anti-logged) coefficient, $\exp(\beta_k)$, represents the expected multiplicative change in the odds of a lower outcome versus a higher outcome, for a unit increase in the predictor. Each coefficient is effectively averaged over all possible combinations of response categories into two groups.

5.4.2 Adjacent Categories Model

Instead of combining response groups as done with the cumulative logit model, the adjacent categories model considers each pair of consecutive outcomes individually. This model incorporates odds ratios comparing each outcome category to the "next" category according to the response order. For example, in the case of predicting the probability of a student performing as advanced, proficient, partially proficient, and unsatisfactory, odds ratios will be calculated comparing advanced to proficient, comparing proficient to partially proficient, and comparing partially proficient to unsatisfactory. Only consecutive groups are directly compared. The model can be written

$$\ln\left(\frac{\pi_{ij}}{\pi_{i,j-1}}\right) = \beta_{0j} + \sum_{k=1}^{K} \beta_k x_{ki} \quad (j = 2, \dots, J), \quad (5.5)$$

where again the intercept is the only coefficient that changes across outcome categories. In these equations the x_{ki} represent observed predictor values while the β_k are unknown effects that must be estimated through the model to predict values for the π_{ij}. Notice the ratio in the logarithm of the adjacent categories model forms the odds of one outcome category (category j) compared to the immediately preceding outcome category (category $j-1$), according to the inherent ordering in the data. Because of the proportional-odds assumption, coefficients can be interpreted generally in terms of odds of general increases or decreases in the observed outcome categories. The anti-logged (or natural exponentiated) coefficient, $\exp(\beta_k)$, represents the expected multiplicative change in the odds of increasing the response by one ordered category, for a unit increase in the predictor. Each coefficient is effectively averaged over all possible combinations of consecutive response categories.

5.4.3 *Continuation Ratio Model*

The continuation ratio model involves forming odds ratios comparing each outcome category to the accumulation of all previous (or following) ordered categories. In the example of predicting probabilities of student outcomes, each level is compared to all lower levels of achievement. The model can be written

$$\ln\left(\frac{\pi_{ij}}{\sum_{l=1}^{j-1} \pi_{il}}\right) = \beta_{0j} + \sum_{k=1}^{K} \beta_k x_{ki} \quad (j = 2, \dots, J), \quad (5.6)$$

where again the intercept is the only coefficient that changes across outcome categories. For this expression the x_{ki} represent observed predictor values while the β_k are unknown effects that must be estimated through the model to predict values for the π_{ij}. Notice the ratio in the logarithm of the continuation ratio model forms the odds of one outcome category (category j) compared to all preceding outcome categories (categories 1 through $j-1$). The anti-logged (or natural exponentiated) coefficient, $\exp(\beta_k)$, represents the expected multiplicative change in the odds of one outcome versus all lesser outcomes, for a unit increase in the predictor. Each coefficient is effectively averaged over all possible response categories.

While the cumulative logit model remains the most common of the ordinal models, both the adjacent categories model and the continuation ratio model can be preferred in specific cases. For example, the adjacent categories model can be preferred in cases in

Table 5.2 *Logistic modeling options for discrete responses.*

Response type	Model option	Interpretations
Binary	Ordinary logistic regression	Odds of success versus failure
Nominal	Multinomial logistic regression	Odds of each outcome category versus baseline category
Ordinal	Cumulative logit multinomial	Odds of greater versus lesser for each outcome category
	Adjacent categories multinomial	Odds of each outcome category versus prior category
	Continuation ratio multinomial	Odds of each category versus all prior categories

which comparisons of consecutive ordered categories are of primary interest, but response categories with greater separation may not be strongly related. The continuation ratio model can be preferred when reaching one response level is the accumulation of all previous levels, such as with education level. Table 5.2 shows modeling options for discrete, categorical responses. Recall that all model options allow for nonnormal responses, nonconstant response variation, and nonlinear relationships with predictors.

5.5 FHS Analysis: Probability of Hypertension

We wish to investigate whether the probability of developing hypertension at any time after baseline in the Framingham Heart Study can be predicted using the following patient properties as predictors.

- Diabetes: diagnosis of diabetes
- Sex: sex, male (1) or female (2)
- Age: age in years
- Cigarettes: number of cigarettes smoked per day
- Cholesterol: total serum cholesterol

To evaluate these relationships, the Framingham Heart Study data were first filtered such that subjects who showed evidence of hypertension at baseline were removed. Therefore those remaining who show evidence of hypertension at any follow-up must have developed symptoms since baseline. After some exploratory data analyses, we will propose an appropriate logistic regression model, evaluate the fit of that model, then proceed to make parameter interpretations and predictions.

5.5.1 Exploratory Data Analyses

Basic exploratory analyses were presented in Chapter 1, but because of the binary nature of the outcome of interest further investigation is required. We first consider contingency tables showing cross-classified counts of hypertension versus the categorical independent variables, as shown in Table 5.3. Descriptively, the values show hypertension to be

Table 5.3 *Framingham Heart Study
observed counts of hypertension by
categorical predictors.*

		Hypertension	
		Yes	No
Total		1 822	1 182
Diabetes	Yes	33	21
	No	1 789	1 161
Sex	Male	773	540
	Female	1 049	642

slightly more prevalent in females than in males, because the proportion of females with hypertension ($1\,049/(1\,049 + 642) \approx 0.62$) is slightly higher than the corresponding proportion for males ($773/(773 + 540) \approx 0.59$). The counts also show that diabetes is rare in the data set, but that hypertension is comparably likely for those with and without diabetes.

There is no evidence of structural zeros (where we would see cross-classifications between the response and predictors without cases in the data set, or counts of zero). Similarly, there is no evidence of issues with complete or quasi-complete separation (where the event of hypertension is completely described by specific classifications of categorical predictors). This is evident by seeing no zeros or even severe imbalance in which some cases show counts of *close to* zero in the contingency tables. Corresponding tables of predictors (such as diabetes by sex) show that there are no relationships among or missing cross-classifications of predictors.

Referring to standard scatter plots of the response (hypertension) versus each of the continuous predictors (age, number of cigarettes per day, and total cholesterol), it is difficult to determine whether there is any strong relationship among variables. Instead it is often informative to produce a logistic histogram plot, in which frequencies of successes and failures are both displayed as in a histogram, but within a single plot.

The logistic histogram plots for age and for total cholesterol are both shown in Figure 5.1, with age used as the predictor in the left panel and total cholesterol used as the predictor in the right panel. Both logistic histogram plots show extensive overlap between observed successes and failures, that is, most values of age and cholesterol have individuals with both successes and failures. This suggests separation will not be a problem for these data. In addition, the simple estimated logistic curve is shown, suggesting an increase in the likelihood of hypertension as both age and cholesterol increase. Of course, the plots are purely descriptive and do not give any indication of significance of a relationship; nor do the respective plots account for additional predictor influence.

5.5.2 Logistic Regression Model

Because the outcome of interest, hypertension, is a binary variable, we will apply a binary logistic regression model to predict the probability of the event of hypertension. Using a model of this type will allow conclusions about the probability of hypertension, about the odds of hypertension, and about the expected change in the odds of hypertension for different

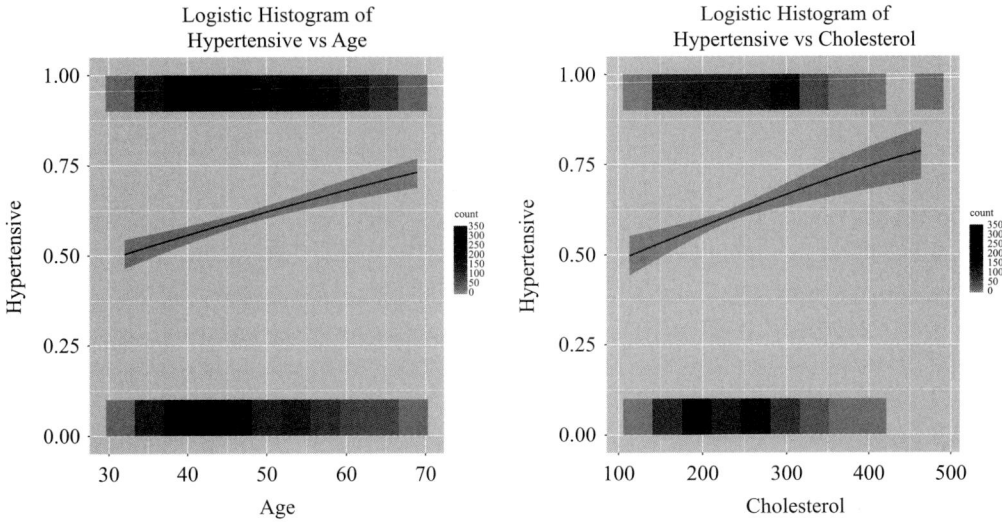

Figure 5.1 Framingham Heart Study logistic histogram plot of hypertension versus age (left panel), logistic histogram plot of hypertension versus tot cholesterol (right panel), darker shading indicates greater numbers.

values of predictors. The continuous predictors (age, number of cigarettes per day, and total cholesterol) will be included as with standard regression models, and the categorical predictors will be included using indicators for diabetes and for sex. In addition, the model will include multiplicative interactions between diabetes and sex, and also between sex and cigarettes per day. These two interactions will allow us to determine whether the effects of both diabetes and also cigarette usage on the probability of hypertension change significantly between males and females. The model can be written in equation form as follows:

$$\ln\left(\frac{\pi_{\text{Hyp}}}{1-\pi_{\text{Hyp}}}\right) = \beta_0 + \beta_1(\text{diabetes}) + \beta_2(\text{sex}) + \beta_3(\text{age}) + \beta_4(\text{cigarettes})$$
$$+ \beta_5(\text{cholesterol}) + \beta_6(\text{diabetes} \times \text{sex}) \qquad (5.7)$$
$$+ \beta_7(\text{sex} \times \text{cigarettes}),$$

where π_{Hyp} indicates the probability of hypertension. Recall that this equation is equivalent to writing the probability of hypertension in terms of all predictors, using a nonlinear relationship,

$$\pi_{\text{Hyp}} = \frac{\exp(\beta_0 + \beta_1(\text{diabetes}) + \cdots + \beta_7(\text{sex} \times \text{cigarettes}))}{1 + \exp(\beta_0 + \beta_1(\text{diabetes}) + \cdots + \beta_7(\text{sex} \times \text{cigarettes}))}.$$

5.5.3 Logistic Regression Model Fit

In evaluating the fit of the model given in Eq. (5.7), the Hosmer-Lemeshow test shows a statistic of $X^2 \approx 14.781$ on 8 degrees of freedom, with associated p-value $p = 0.065$. The p-value is relatively large, and therefore we fail to reject the null hypothesis of good fit and conclude that the data fail to provide evidence of a poor fit.

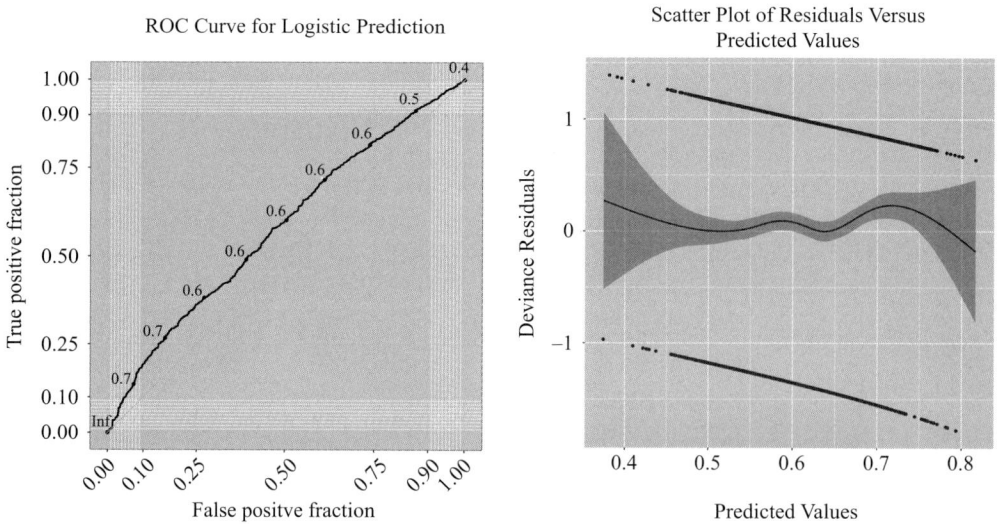

Figure 5.2 Framingham Heart Study hypertension logistic regression model ROC curve (left panel), deviance residuals versus predicted values (right panel).

Turning next to the ROC curve, the plot in the left panel of Figure 5.2 shows the anticipated concave shape, although the curve does not deviate very far from the diagonal. This plot tells us that any increase in sensitivity (the ability to correctly predict hypertension, the vertical height in the plot) is accompanied by an almost equivalent loss in specificity (the ability to predict lack of hypertension, the horizontal distance in the plot). In addition, the area under the ROC curve is estimated to be 0.5796, which supports the mediocre predictive capability of the model. (Remember that all classifications should provide an area of at least 0.50, with values rarely exceeding 0.85 or 0.90 before the classification is so accurate that the model fails to converge.) Neither the Hosmer-Lemeshow test nor the ROC curve provide evidence of a poor fit, although the accuracy of predictions is not strong.

Next considering the standardized deviance residuals, we take values exceeding 3.0 standard deviations from the mean to be evidence from the data that the model does not fit well. Logistic regression residual plots often show patterns because of the nature of the calculation of deviance residuals using raw residuals based on responses restricted to 0 or 1, and therefore the deviance residual plot in the right panel of Figure 5.2 is expected to show patterns and will be used only to identify extreme values. None of the residuals exceeds 2.0 in absolute magnitude, with the largest deviance residual equal to -1.7766, and therefore the residuals fail to suggest the model fits poorly.

Considering the deviance statistic overall, the residual deviance is approximately $3\,892.2$ on $2\,938$ degrees of freedom, giving a ratio of $3\,892.2/2\,938 \approx 1.325$. This ratio is not much larger than 1, and therefore fails to provide evidence of a poor fit to the model. Data analysts often turn to a measure of pseudo R^2, such as McFadden's pseudo R^2 (which is approximately 0.0335 here). As expected, the pseudo R^2 value is quite small, and we prefer to use the other fit statistics in evaluating the model. Overall none of the standard diagnostics suggest the model fits the data poorly, but the predictive ability does not appear to be strong. This is likely due to missing predictors that could help prediction accuracy.

Table 5.4 *Framingham Heart Study results of logistic regression of hypertension.*

Predictor	Estimate	Std. error	z-value	p-value
Intercept	−1.242	0.291	−4.277	< 0.001
Diabetes	−0.303	0.354	−0.856	0.392
Sex (female)	0.191	0.104	1.831	0.067
Age	0.022	0.005	4.413	< 0.001
Cigarettes	−0.004	0.004	−0.852	0.394
Cholesterol	0.003	0.001	2.740	0.006
Diabetes × sex	0.749	0.634	1.180	0.238
Sex × cigarettes	−0.013	0.007	−1.867	0.062

Residual deviance	3892.15 on 2938 d.f.
Null deviance	3948.25 on 2945 d.f.
AIC	3908.2

Table 5.5 *Framingham Heart Study estimated odds ratios for hypertension, with 95% confidence intervals.*

Predictor	Odds ratio	95% Confidence interval
Diabetes	0.738	(0.369,1.500)
Sex (female)	1.210	(0.986,1.484)
Age	1.022	(1.012,1.032)
Cigarettes	0.996	(0.988,1.005)
Cholesterol	1.003	(1.001,1.004)
Diabetes × sex	2.114	(0.633,7.885)
Sex × cigarettes	0.987	(0.974,1.001)

5.5.4 Model Parameter Interpretations

The results of fitting the logistic regression model can be found in Table 5.4. Predictor significance is determined by Wald tests, whose p-values are provided in the final column of the table. Coefficients can be interpreted very efficiently as generally indicating expected increases or decreases in the probability of hypertension. For example, the coefficient for the age predictor is 0.022, which is positive and significant. The data show that older populations are expected to have a significantly higher probability of showing hypertension. Often these types of interpretations are sufficient to understand the effects of predictors in the model.

We can more rigorously quantify predictor effects by interpreting coefficients using odds ratios, where the ratio is interpreted as the odds of hypertension versus no hypertension, shown in Table 5.5. For age, $\exp(0.022) \approx 1.022$ indicates that the odds of hypertension is expected to increase by a multiple of 1.022 (or increase by $(1.022 - 1) \times 100\% = 2.2\%$) for each increase of one year of age. The associated 95% confidence interval for the odds ratio is estimated as $(1.012, 1.032)$, indicating that, with 95% confidence, the true odds ratio is expected to be within the range from 1.012 to 1.032. For logistic regression predictors, two-sided significance according to the p-value is equivalent to confidence intervals that do not contain the value 1, as is true for age here.

Note that the interpretation of the odds ratio for age does *not* give any indication of what happens as an individual ages. To obtain such an interpretation it is necessary to collect data from the same individuals over time. Instead this odds ratio represents a comparison between two populations, similar with respect to diabetes, sex, cigarette use, and total cholesterol, but with one population older than the other by an average of one year. Specifically, for a population that is on average one year older than another similar population, the odds of hypertension is expected to be higher by a multiple of 1.022 for the older population. Equivalently, the odds of hypertension is expected to increase by 2.2% for older populations. For greater differences, such as a 10-year difference in average age, $\exp(10 \times 0.022) \approx 1.246$ provides the odds ratio. Notice this is not simply ten times the original odds ratio, as the relationship between age and odds ratio is not linear.

For the marginally significant sex indicator for females, $\exp(0.191) \approx 1.210$ (with associated 95% confidence interval $(0.986, 1.484)$) indicates that the odds of hypertension increases by a multiple of 1.210 (or by 21%) for females, as compared to similar males. For example, for a population of males with odds of hypertension given by, say, 3.00, a similar population of females would be expected to have an odds of hypertension of $1.210 \times 3.00 \approx 3.630$. In general females are expected to have a higher odds of hypertension than males. Notice the 95% confidence interval of the odds ratio for sex contains 1, which corresponds to the nonsignificance of sex at the 0.05 significance level.

The only other significant predictor is total cholesterol, which reveals that the odds of hypertension is expected to increase by a factor of $\exp(0.003) \approx 1.003$ (with associated 95% confidence interval $(1.001, 1.004)$) for a unit increase in cholesterol, all other variables remaining unchanged. For an increase in cholesterol of 25, the odds of hypertension is expected to increase by a factor of $\exp(25 \times 0.003) \approx 1.078$, or an increase of 7.8%. In general higher cholesterol is associated with a higher expected probability of hypertension.

Based on Wald tests, the interaction effects do not show significance at the $0.05/7 \approx 0.007$ significance level (conservatively adjusting for the number of tests performed using the same data). The coefficient for the diabetes $\times$ sex interaction, if significant, would represent a meaningful adjustment to the effect of diabetes for females. If significant, the coefficient for the sex and cigarettes per day interaction would represent a change in the effect of cigarettes per day for females. While the interactions cannot be reported as significant, they should be used in making predictions because they have been included in the model. Recall that all predictors included in a model can help improve the robustness of resulting predictions.

5.5.5 Model Prediction

Fitting of the logistic regression model produces the following predicted regression function:

$$\ln\left(\frac{\hat{\pi}_{\text{Hyp}}}{1-\hat{\pi}_{\text{Hyp}}}\right) = -1.242 - 0.303(\text{diabetes}) + 0.191(\text{sex}) + 0.022(\text{age})$$
$$- 0.004(\text{cigarettes}) + 0.003(\text{cholesterol}) \tag{5.8}$$
$$+ 0.749(\text{diabetes} \times \text{sex}) - 0.013(\text{sex} \times \text{cigarettes}).$$

Table 5.6 *Framingham Heart Study predicted probabilities of hypertension for diabetic and non-diabetic males and females, ages 40 through 70.*

		Age						
		40	45	50	55	60	65	70
Males	Non-diabetic	0.552	0.578	0.605	0.632	0.656	0.681	0.704
	Diabetic	0.476	0.504	0.531	0.558	0.585	0.612	0.637
Females	Non-diabetic	0.568	0.595	0.621	0.647	0.671	0.695	0.718
	Diabetic	0.672	0.696	0.719	0.741	0.761	0.781	0.799

Given values of all predictors, an individual's probability of developing evidence of hypertension can be predicted using the inverse of the logit transformation,

$$\hat{\pi}_{\text{Hyp}} = \frac{\exp(-1.242 - 0.303(\text{diabetes}) + \cdots - 0.013(\text{sex} \times \text{cigarettes}))}{1 + \exp(-1.242 - 0.303(\text{diabetes}) + \cdots - 0.013(\text{sex} \times \text{cigarettes}))}. \quad (5.9)$$

For the purposes of prediction, predictor significance is not considered. That is, all predictors in the model are used to make predictions of the probability of interest. Typically, predicted probabilities of successful outcome are not reported for all possible values of predictors, or even for all observed values of predictors, but rather for an informative subset of values. These predictor values should be selected to be on the interior of the range of values observed; that is, predictions should only be made within the ranges of values of predictors observed through data collection.

Table 5.6 shows the predicted probabilities of developing hypertension, for both males and females with and without diabetes, from age 40 to age 70. Prediction has been made using the average of all observed values for both the number of cigarettes used per day, and the total cholesterol, as any model prediction requires specific values of all predictors. We see that the predicted values clearly show higher expectations of hypertension for females than for males. These conclusions are consistent with the exploratory analyses performed prior to fitting the model, but now include predictions that account for other variables in the model.

The predicted values also show non-diabetic males to have a higher expected likelihood of developing hypertension, while the opposite is true for females. This interaction between sex and diabetes is interesting, but only descriptive because it was not found to be significant in the model. Predicted values also allow for very specific conclusions to be stated, such as: "It is expected that around 75% of diabetic females aged 60 will show evidence of hypertension."

It is often valuable to visualize the magnitudes of predicted probabilities using a predicted value scatter plot and smooth curve. Figure 5.3 shows a smooth predicted probability curve across values of age (on the horizontal) for each combination of sex and diabetes. Also included on the plots are the observed values of hypertension (either 0 or 1) and shaded regions indicating two standard errors greater and less than the predicted probability curve. Notice the standard errors are much greater for the two diabetes groups, as the smaller sample of individuals with diabetes in the data set corresponds to less confidence in predictions.

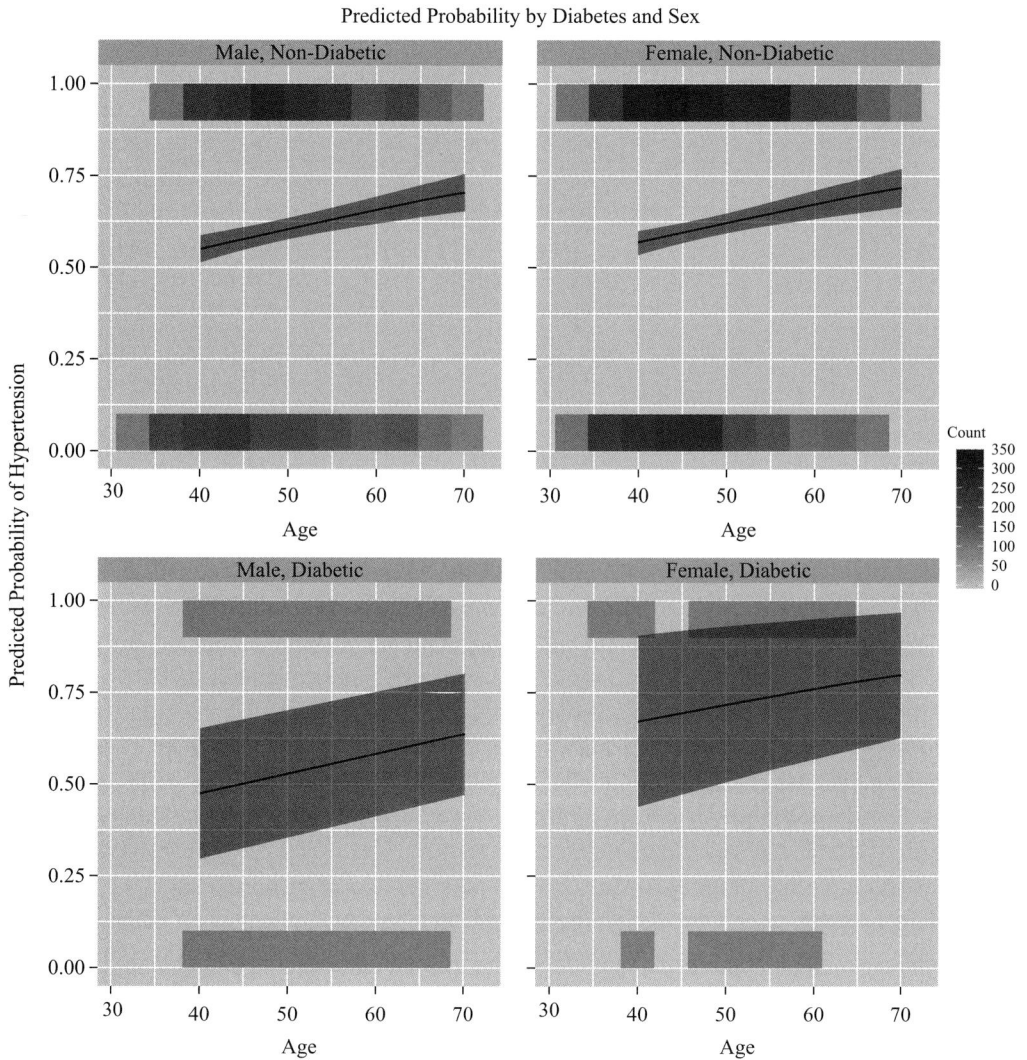

Figure 5.3 Framingham Heart Study scatter plot of hypertension versus age with smooth prediction, for non-diabetic males (first panel), non-diabetic females (second panel), diabetic males (third panel), diabetic females (fourth panel).

We applied a logistic regression model to accommodate the binary nature of the outcome of interest, hypertension. The nonlinear logistic model produces predicted probabilities within the range of reasonable values between 0 and 1, regardless of the values of predictors. From the analysis we found that sex, diabetic status, age, cigarette use, and total cholesterol can be used to effectively predict the probability of hypertension. Females, older individuals, and those with higher total cholesterol show an increase in the likelihood of hypertension. There does not appear to be a significant relationship between either diabetes or cigarette use and the chance of hypertension, after controlling for the other variables in the study.

5.6 SSOCS Analysis: Probability of Bullying

Using the School Survey on Crime and Safety data, we would like to determine the factors that impact school bullying, including the predictors listed below.

- Uniforms: whether the school uses standard uniforms
- Metal detectors: whether students pass through metal detectors
- Tipline: whether the school has a "tipline" to report issues
- Counseling: whether the school offers student counseling
- Suspensions: the number of suspensions due to insubordination
- Crime: a measure of crime in the area of the school (low, moderate, high)

Each school's response for the outcome of interest, bullying, represented a category that best described the school's bullying frequency: never, on occasion, monthly, weekly, or daily. Because the response is represented by categories, we will model the probabilities associated with all five self-reported categories.

This type of outcome can be viewed as an ordinal response because the self-reported identifiers represent incremental increases in bullying frequency. Daily bullying is a "greater" amount of bullying than weekly bullying, weekly bullying is a "greater" amount of bullying than monthly bullying, and so on. Because of this property of the outcome of interest, we can perform more powerful tests by comparing response categories in such a way as to take advantage of the inherent ordering in the values. Specifically, we will model probabilities associated with each of the five response groups using the adjacent categories model. The adjacent categories model will produce odds ratios interpreted as related to an expected change from one bullying category to the next. Any of the multinomial models will provide appropriate interpretations for these data, but we chose the incremental change interpretation provided by the adjacent categories model because actual changes in school environment can reasonably be seen as happening in increments, i.e., changing from monthly bullying to weekly bullying.

After presenting some exploratory data analyses, we will propose an appropriate ordinal multinomial model. We will then consider the fit of the model, followed by parameter interpretations and model-based predictions.

5.6.1 Exploratory Data Analysis

Some exploration of the data was performed in Chapter 1. Additional descriptive statistics pertaining to categorical variables are presented here to help inform the most appropriate model. Table 5.7 revisits the counts (for categorical variables) and means and variances (for continuous variables) for each level of bullying.

The values in the table give us a general impression of totals and trends across bullying groups. The most prevalent bullying group is "on occasion," while the least prevalent group is "never." The left panel of Figure 5.5 shows a stacked bar plot of counseling for different levels of bullying, and shows the number of schools with counseling to increase from "never" through "weekly" bullying, but the pattern is unclear for the final level of "daily" bullying because there are far fewer schools in this highest category. Using the counts from the table, the odds of counseling clearly changes across bullying groups, as the "never"

Table 5.7 *School Survey on Crime and Safety descriptive statistics of categorical and continuous predictors by bullying level.*

		Bullying				
		Never	On occasion	Monthly	Weekly	Daily
Total		40	1 187	547	498	288
Uniforms	Yes	10	174	83	66	44
	No	30	1 013	464	432	244
Metal detectors	Yes	1	35	10	12	7
	No	39	1 152	537	486	281
Tipline	Yes	12	405	190	184	110
	No	28	782	357	314	178
Counseling	Yes	30	1 115	509	476	276
	No	10	72	38	22	12
Crime	Low	31	938	416	350	187
	Moderate	6	189	111	117	71
	High	3	60	20	31	30

	Bullying				
	Never	On occasion	Monthly	Weekly	Daily
Total suspensions	40	1 187	547	498	288

group has an odds of counseling of $30/10 = 3.00$, while the "on occasion" group increases to $1\,115/72 \approx 15.49$, and also ≈ 13.39, ≈ 21.64, and ≈ 23.00 for "monthly," "weekly," and "daily," respectively. Even though the odds of counseling is not the outcome of interest, such calculations give us the impression that the availability of counseling and level of bullying are related. Figure 5.5 shows a plot of these odds in the right panel, clearly showing the increasing relationship.

Investigating the counts of bullying levels for different crime levels shows that the most common bullying level is "on occasion" for all crime levels. However, the proportion of schools reporting "daily" bullying increases across crime level, from $187/(31+938+416+350+187) \approx 0.097$ for schools in areas of low crime, to ≈ 0.144 and ≈ 0.208 for schools in areas of moderate and high crime, respectively. Changes in these proportions suggest an association between area crime and school bullying level. Figure 5.4 shows proportions associated with each bullying level, connected for each level of area crime. The plot makes it clear that the proportions for higher levels of bullying increase with area crime.

5.6.2 Ordinal Multinomial Model

Because the outcome of interest, level of school bullying, is an ordinal categorical variable, we will apply an ordinal multinomial logistic regression model to predict the probability of any school reporting each of the bullying levels. Specifically, we will apply the adjacent categories model in order to easily report odds ratios comparing consecutive levels of bullying. As a general interpretation, we will be able to report the change in the odds of moving up one level to a higher bullying category according to each of the predictors: the

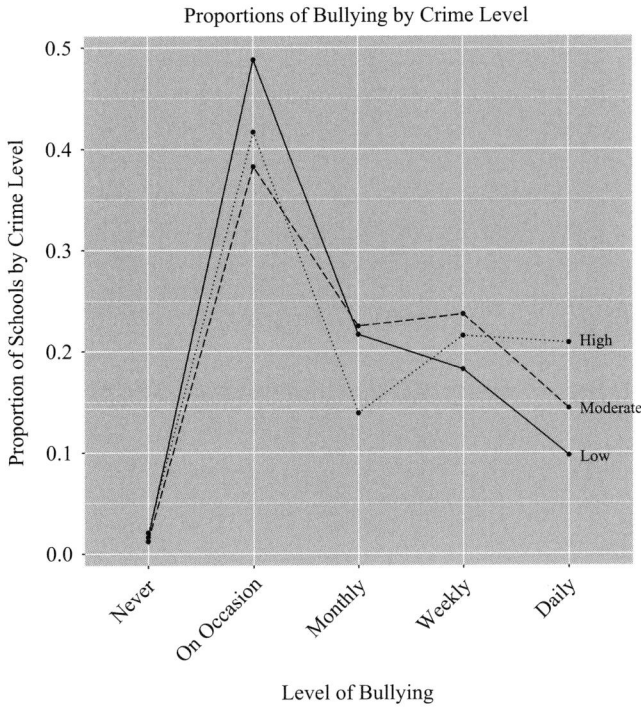

Figure 5.4 School Survey on Crime and Safety proportions of schools with different levels of bullying, by area crime level.

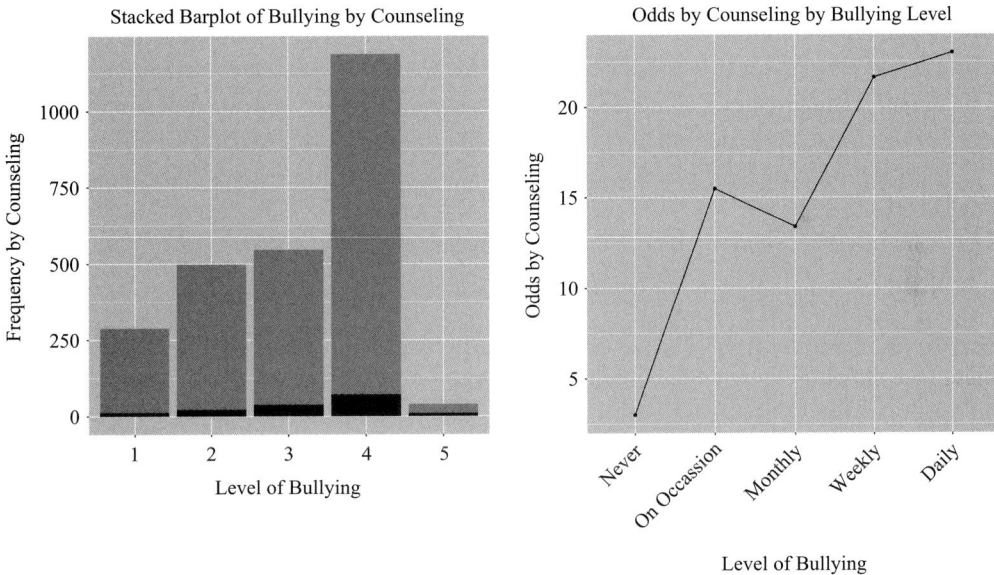

Figure 5.5 School Survey on Crime and Safety stacked bar plot of frequency of bullying levels, light shading for counseling and dark shading for no counseling (left panel), line plot of odds of counseling by bullying level (right panel).

school's uniform policy, presence of metal detectors, existence of a tipline, availability of counseling, the number of suspensions, and the school location's crime level (low, moderate, high). In addition, our model will include an interaction effect between the number of suspensions and area crime level. This interaction effect will allow us to determine whether the impact of the number of suspensions on the probability of reporting different levels of bullying depends on the school location crime level. Finally, we will assume a proportional odds model, implying that the effect of each predictor on the odds of increasing to the next level of bullying is the same for all levels of bullying. The model can be written using of four separate logistic regression equations, one associated with each comparison of consecutive levels of school bullying,

$$\ln\left(\frac{\pi_o}{\pi_n}\right) = \beta_{01} + \sum_{k=1}^{9} \beta_k x_k,$$

$$\ln\left(\frac{\pi_m}{\pi_o}\right) = \beta_{02} + \sum_{k=1}^{9} \beta_k x_k,$$

$$\ln\left(\frac{\pi_w}{\pi_m}\right) = \beta_{03} + \sum_{k=1}^{9} \beta_k x_k,$$

$$\ln\left(\frac{\pi_d}{\pi_w}\right) = \beta_{04} + \sum_{k=1}^{9} \beta_k x_k, \tag{5.10}$$

where π_n indicates the probability that a school will respond "never" to the bullying question, π_o is the probability of reporting "on occasion," π_m is the probability of reporting "monthly," π_w the probability of reporting "weekly," and π_d the probability of "daily" bullying being reported. The intercepts β_{0j} are allowed to change across outcome comparisons, but the rest of the predictor effects remain identical, which reflects the proportional-odds assumption. Thus, the final term in each equation remains the same:

$$\sum_{k=1}^{9} \beta_k x_k = \beta_1(\text{uniforms}) + \beta_2(\text{metal detectors}) + \beta_3(\text{tipline}) + \beta_4(\text{counseling})$$

$$+ \beta_5(\text{suspensions}) + \beta_6(\text{moderate crime}) + \beta_7(\text{high crime}) \tag{5.11}$$

$$+ \beta_8(\text{suspensions} \times \text{moderate crime}) + \beta_9(\text{suspensions} \times \text{high crime}),$$

where "low" crime is taken as the reference group to which the other levels of area crime will be compared. Notice the expression on the left-hand side of each model equation represents a comparison of one bullying category to the level immediately preceding, in terms of extent of bullying. Prediction functions will be discussed in detail in the final subsection of this analysis, as multinomial models are typically not written as explicit formulas for each probability of interest, which can be done easily for the binary logistic regression model.

5.6.3 Ordinal Multinomial Model Fit

The statistics available for evaluating the fit of a multinomial logistic regression model are not as extensive as those used with ordinary logistic regression models. We can consider the

Table 5.8 *School Survey on Crime and Safety adjacent categories multinomial logistic regression residuals for each outcome model.*

Event modeled	Residual type	Minimum	Median	Maximum
On occasion	Pearson	−10.362	0.036	0.291
	Deviance	−109.627	−3e-5	7.265
Monthly	Pearson	−2.285	0.348	1.698
	Deviance	−7.265	0.000	5.128
Weekly	Pearson	−1.441	−0.309	2.293
	Deviance	−5.128	0.000	8.125
Daily	Pearson	−1.297	−0.110	4.214
	Deviance	−8.125	0.000	21.081

ratio of the residual deviance to the residual degrees of freedom as an indication of fit, which in this case is $6683.915/10227 \approx 0.654$. Values much greater than 1 can be an indication of poor fit, which is not the case here.

Both Pearson residuals and deviance residuals can be used to investigate the fit of multinomial logistic regression models. Remember that for logistic models, residuals are used to identify unusually large values, indicating poor fit. Table 5.8 shows the Pearson and (unstandardized) deviance residuals associated with each component of the adjacent-categories model. We can see that there appears to be unusually large negative residuals for the model comparing "on occasion" to "never," and there appear to be some large positive residuals for the model comparing "daily" to "weekly" bullying. The negative residuals indicate that the expectation of the event is much greater than actually observed, while positive residuals suggest an expectation of the event that is less than observed. Overall these residuals tell us that the model appears to fit well for moderate categories of bullying, but the fit suffers for the more extreme levels of bullying, such as when the model involves either "never" or "daily" bullying.

The proportional-odds assumption is typically evaluated by performing a χ^2 test comparing model deviance to a similar model allowing for nonproportional odds. For these data, the nonproportional odds adjacent categories model fails to converge, suggesting the proportional-odds assumption is necessity to fit the model.

5.6.4 Model Parameters Interpretations

The results of fitting the adjacent categories multinomial logistic regression model are presented in Table 5.9. The final column shows the *p*-values associated with Wald tests, which are used to determine significance of each parameter. First we interpret the intercepts associated with each outcome comparison, all significant according to the Wald tests. The natural exponential of the values can be interpreted loosely as a "baseline" odds associated with an increase in the level of bullying. For example, for the model of "on occasion" versus "never" the intercept gives $\exp(3.156) \approx 23.477$, which is an estimate of the odds of occasional bullying versus no bullying for schools with no uniform policy, no metal detectors, no tipline, no counseling, zero suspensions, and in an area of low crime. The

Table 5.9 *School Survey on Crime and Safety results of adjacent categories multinomial logistic regression of bullying.*

Predictor	Estimate	Std. error	z-value	$Pr(>|z|)$
Intercept (occasion versus never)	3.156	0.178	17.721	< 0.001
Intercept (monthly versus occasion)	−1.024	0.095	−10.805	< 0.001
Intercept (weekly versus monthly)	−0.359	0.103	−3.501	< 0.001
Intercept (daily versus weekly)	−0.831	0.112	−7.429	< 0.001
Uniforms	−0.112	0.056	−1.985	0.047
Metal detectors	−0.198	0.126	−1.576	0.115
Tipline	0.051	0.039	1.308	0.191
Counseling	0.204	0.084	2.430	0.015
Suspensions	0.0004	0.0003	1.152	0.249
Moderate crime	0.241	0.048	5.010	< 0.001
High crime	0.299	0.083	3.596	0.0003
Suspensions × moderate crime	0.0003	0.0007	0.386	0.699
Suspensions × high crime	0.001	0.001	0.840	0.401

Residual deviance 6 683.915 on 10 227 d.f.
Log-likelihood −3 341.957 on 10 227 d.f.

Table 5.10 *School Survey on Crime and Safety estimated odds ratios of increased bullying with 95% confidence intervals.*

Coefficient	Odds ratio	95% Confidence interval
Uniforms	0.894	(0.801,0.999)
Metal detectors	0.820	(0.641,1.049)
Tipline	1.052	(0.975,1.135)
Counseling	1.227	(1.040,1.447)
Suspensions	1.000	(0.999,1.001)
Moderate crime	1.273	(1.158,1.399)
High crime	1.348	(1.146,1.587)
Suspensions × moderate crime	1.000	(0.999,1.002)
Suspensions × high crime	1.001	(0.998,1.004)

odds value is largest for the lowest bullying group comparison ("on occasion" versus "never") because these groups have the greatest opportunity to increase, and because the "on occasion" group has the largest sample size and therefore the largest baseline likelihood.

Both crime indicators show clear significance within the model. This tells us that the odds of an increase in bullying is expected to change significantly between schools in areas with different levels of crime. Specifically, the coefficient for moderate crime is 0.241, with odds ratio exp(0.241) ≈ 1.273. This means that the odds of an increase in reported bullying level for a school in an area of moderate crime is expected to be 1.273 times that of a similar school in an area of low crime, or an (1 − 1.273) × 100% = 27.3% increase. The odds of an increase in reported bullying level for a school in an area of high crime is expected to be exp(0.299) ≈ 1.348 times that of a similar school in an area of low crime, or an increase of 34.8%. Odds ratios and associated 95% confidence intervals are displayed for all parameters except intercepts in Table 5.10.

Notice the interpretation for crime is made generally for all changes: this odds ratio is predicted to be the same for a move from "never" to "on occasion" as for a move from "on occasion" to "monthly," which is the same odds ratio as for a move from "monthly" to "weekly," and so on. This is the consequence of the proportional-odds assumption, that the likelihood of change is equivalent for all groups. Also notice that the interpretation is made as an *increase* in bullying level, and not just a change from one outcome group to another. This is a consequence of the adjacent categories model taking advantage of the ordinal nature of the data, and making a specific and statistically powerful comparison in one direction.

Results show that the counseling effect is significant ($p = 0.015$), with parameter estimate 0.204. Schools with counseling services available to students are expected to have an increase of a factor of $\exp(0.204) \approx 1.227$ in the odds of increased bullying, as compared to similar schools without counseling. Such a result should not be taken as a causal conclusion: offering counseling services does not necessarily increase the likelihood of bullying. Rather, it is more likely that schools with higher levels of bullying tend to offer counseling services for students.

Table 5.9 shows a marginally significant uniform policy effect ($p = 0.047$), with parameter estimate -0.112. Schools with a uniform policy are expected to have a decrease in the odds of increased bullying, by a factor of $\exp(-0.112) \approx 0.894$, as compared to similar schools without a uniform policy. Often it is preferred to reverse conclusions for negative parameter estimates within a logistic model. Schools *without* a uniform policy are expected to show an increase by a factor of $\exp(0.112) \approx 1.119$ in the odds of increased bullying, as compared to similar schools with uniform policies.

Often general directional conclusions are of interest to the researchers. Based on this analysis, higher crime in the area where a school is located is associated with increased odds of higher levels of bullying, controlling for uniform policies, availability of counseling, and all of the other predictors. Availability of counseling for students is associated with greater odds of increased levels of bullying, although this conclusion may represent the reverse causality of increased levels of bullying leading to counseling services. Having a uniform policy is associated with a decrease in the odds of higher levels of bullying, accounting for all other predictors.

5.6.5 Model Prediction

The prediction equation for ordinal multinomial logistic regression models involves more terms than for ordinary binary logistic regression model prediction. Using the notation $\text{RHS}_j = \beta_{0j} + \sum_{k=1}^{9} \beta_k x_k$ for the "right-hand side" of each model equation from the expressions in Equation 5.10, predicted probabilities for each group can be written

$$\pi_n = \frac{1}{\text{total factor}}, \qquad \pi_o = \frac{\exp(\text{RHS}_1)}{\text{total factor}},$$

$$\pi_m = \frac{\exp(\text{RHS}_1 + \text{RHS}_2)}{\text{total factor}}, \qquad \pi_w = \frac{\exp(\text{RHS}_1 + \text{RHS}_2 + \text{RHS}_3)}{\text{total factor}},$$

$$\pi_d = \frac{\exp(\text{RHS}_1 + \text{RHS}_2 + \text{RHS}_3 + \text{RHS}_4)}{\text{total factor}}, \tag{5.12}$$

where the "total factor" in the denominator is a combination of components of the original adjacent-categories model equations,

$$\text{total factor} = 1 + \exp(\text{RHS}_1) + \exp(\text{RHS}_1 + \text{RHS}_2)$$
$$+ \exp(\text{RHS}_1 + \text{RHS}_2 + \text{RHS}_3)$$
$$+ \exp(\text{RHS}_1 + \text{RHS}_2 + \text{RHS}_3 + \text{RHS}_4). \tag{5.13}$$

Table 5.11 displays the minimum, median, and maximum predicted probabilities for each level of bullying, based on observed data. It is interesting that the minimum probability is highest for weekly bullying, but the median and maximum of predicted probabilities both peak for occasional bullying. This implies that all schools have a relatively high floor for the likelihood of weekly bullying, but the greatest chances are associated with reporting occasional bullying.

Also shown in Table 5.11 are the predicted probabilities of each level of bullying for schools with a uniform policy, no metal detectors, no tipline, counseling services, the average number of suspensions observed in the data, for all of low, moderate, and high crime locations. For such schools in areas of low crime, our model predicts bullying on occasion to have the highest likelihood ($\hat{\pi}_o \approx 0.531$), followed by monthly bullying as the next most likely ($\hat{\pi}_m \approx 0.210$). For similar schools in areas of moderate crime, occasional bullying remains the most likely ($\hat{\pi}_o \approx 0.429$), but weekly bullying is almost as likely as monthly bullying, with both at a predicted probability of around 0.21. Considering similar schools in areas of high crime, weekly bullying does indeed show as the second most likely outcome ($\hat{\pi}_w \approx 0.226$). These predictions tell us that, as the crime in the area of the school increases, schools are less likely to report occasional bullying and more likely to report weekly bullying. Figure 5.6 shows these predicted probabilities visually, supporting the conclusions from the table.

Table 5.11 *School Survey on Crime and Safety predicted probabilities of bullying using observed predictor values.*

	Level of bullying				
Predicted group	Never	On occasion	Monthly	Weekly	Daily
Minimum observed	0.001	0.067	0.101	0.122	0.047
Median observed	0.017	0.486	0.215	0.184	0.099
Maximum observed	0.029	0.607	0.216	0.309	0.538
Low crime	0.021	0.531	0.210	0.161	0.077
Moderate crime	0.013	0.429	0.216	0.212	0.130
High crime	0.011	0.400	0.216	0.226	0.147

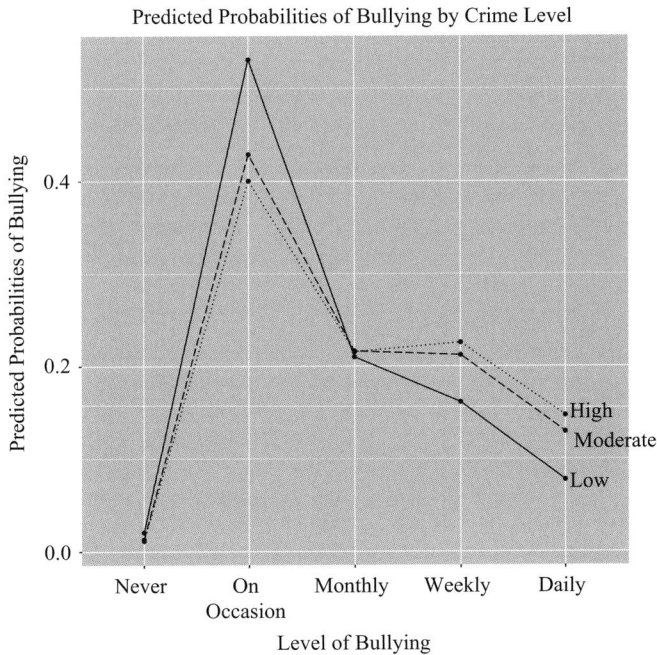

Figure 5.6 School Survey on Crime and Safety predicted probabilities of levels of bullying, by area crime level.

We found the adjacent categories ordinal multinomial model to be an appropriate choice for the ordered, multi-category response of bullying level. This nonlinear logistic model allowed us to predict the probability associated with each level of bullying, and to report odds comparing each level to the next lower level of bullying. The analysis showed us that the odds of greater bullying increases for schools in areas with higher crime, for schools with counseling services available to students, and for schools without uniforms.

5.7 Clickstream Analysis: Probability of Redlink

The English Wikipedia data contain records of the title of the previous page visited, prior to Wikipedia, along with records of whether the requested page was not available, also referred to as a "redlink." We would like to predict the probability of producing a redlink after redirection through Wikipedia, during the first approximately 3 hours of February 2015, using the previous title as the only predictor.

• Previous title: the previous page title category

We will present some exploratory analyses, followed by an appropriate logistic regression model. Next we will assess the fit of the model, followed by interpretations of model coefficients and model predictions.

Table 5.12   *English Wikipedia Clickstream counts of observed redlinks by previous page title groups.*

Previous page	Redlink	No redlink
Total	317	65 150
Bing	0	1 593
Google	1	8 077
Yahoo	0	1 231
Wikipedia	0	3 961
Main_Page	0	509
Empty	0	6 733
Other	316	43 046

5.7.1 Exploratory Data Analysis

We begin by exploring the outcome of interest, instance of a redlink, and its association with the predictor of interest, the previous title group. Previous title was grouped by common sites that direct to Wikipedia, and includes a Bing page, a Google page, a Yahoo page, another Wikipedia page, the link's main page on Wikipedia, a group for visits that did not come from another page (listed as "empty"), and a general group, "other," that does not fit any of the other classifications. Table 5.12 shows the number of redlinks associated with each unique previous page.

Recall that the English Wikipedia data contain counts of unique pairs of previous and requested titles, meaning the data in Table 5.12 represent counts of unique pairs of previous pages leading to a redlink. For example, the count of 1 for Google means that there is a single pair consisting of a Google page and Wikipedia request leading to a redlink. This single pair may have been observed many times during the window represented by the data, but multiple instances of the same pair are not indicated by this table. That information is contained in the count variable.

We can make a number of interesting conclusions based on this table. The redlink is a rare event, with only about $317/65\,467 \times 100\% \approx 0.484\%$ of pairs leading to a redlink. Only Google and "other" previous pages led to a redlink during the time under consideration, however, we can still believe that there is some probability that the other previous title sources could lead to a redlink under additional data collection. The information in Table 5.12 is represented visually in Figure 5.7, showing a logistic bar plot in which darker shades represent a greater number of pairs of previous title-redlink combinations.

5.7.2 Logistic Regression Model

Because the outcome of interest, redlink, is a binary variable with only two possible outcomes, we choose to apply an ordinary logistic regression model. Using "other" as the reference category to which all other previous title groups will be compared, the model can be written

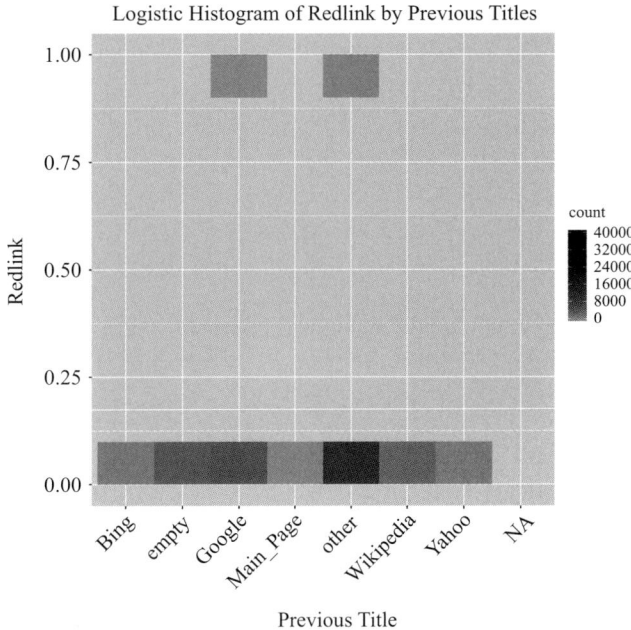

Figure 5.7 English Wikipedia Clickstream logistic histogram plot of redlink versus previous title, darker shading indicates greater numbers of redlinks.

$$\ln\left(\frac{\pi_r}{1-\pi_r}\right) = \beta_0 + \beta_1(\text{Bing}) + \beta_2(\text{Google}) + \beta_3(\text{Yahoo}) + \beta_4(\text{Wikipedia})$$
$$+ \beta_5(\text{Main\_Page}) + \beta_6(\text{empty}), \tag{5.14}$$

where π_r indicates the probability of a redlink. Because the data are saved as unique pairs of previous title groups and redlink indicators, with a separate variable indicating the number of each unique pair, we will apply a weighted estimation using the variable indicating the counts as weights. Recall that this model can be written equivalently in terms of the probability of interest

$$\pi_r = \frac{e^{(\beta_0 + \beta_1(\text{Bing}) + \beta_2(\text{Google}) + \beta_3(\text{Yahoo}) + \beta_4(\text{Wikipedia}) + \beta_5(\text{Main\_Page}) + \beta_6(\text{empty}))}}{1 + e^{(\beta_0 + \beta_1(\text{Bing}) + \beta_2(\text{Google}) + \beta_3(\text{Yahoo}) + \beta_4(\text{Wikipedia}) + \beta_5(\text{Main\_Page}) + \beta_6(\text{empty}))}}. \tag{5.15}$$

5.7.3 Logistic Regression Model Fit

The fit of a logistic regression model is traditionally assessed, in part, by applying the Hosmer-Lemeshow test. However, the Hosmer-Lemeshow test is known to be fundamentally overpowered when sample sizes exceed even 25 000 cases. That is, the test will almost certainly show significance and suggest poor fit regardless of the fit of the model to the data. In this case the expectation holds, as the Hosmer-Lemeshow test shows a test statistic of $X^2 = 274.75$ with a p-value smaller than any plausible significance level. Instead of relying on this test, we will consider deviance residuals and predictive classification more strongly.

104 *Discrete, Categorical Response Models*

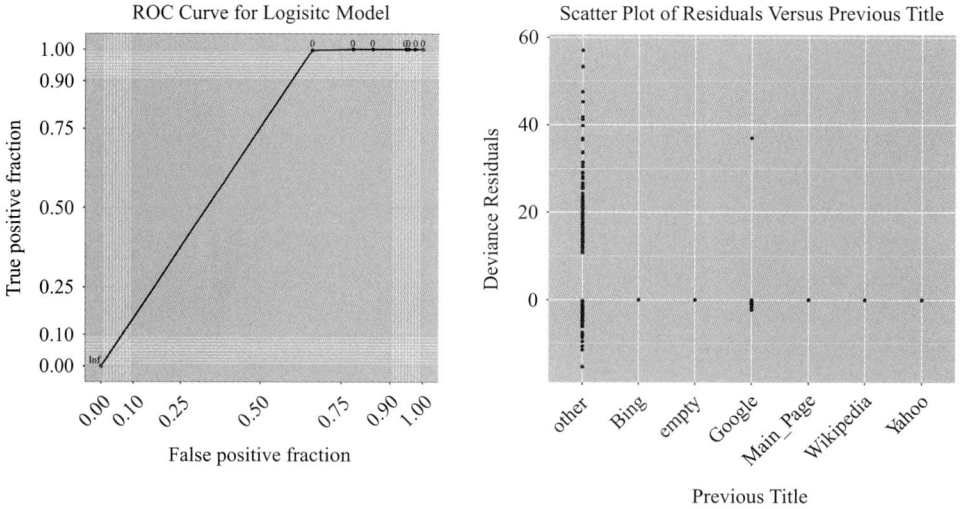

Figure 5.8 English Wikipedia Clickstream redlink logistic regression model ROC curve (left panel), deviance residuals versus predicted values (right panel).

The overall residual deviance is 114 127 on 65 460 degrees of freedom, giving a ratio of $114\,127/65\,460 \approx 1.743$, which does not provide evidence of general poor fit. The left panel of Figure 5.8 shows the ROC curve for the model, with associated area under the curve of 0.668. This area shows acceptable predictive ability, although the ROC curve itself does not show as much of an increase in sensitivity without loss of specificity as would be ideal. In other words, the curve stays too close to the diagonal line, which represents random guessing.

The right panel of Figure 5.8 shows a plot of the standardized deviance residuals versus the previous title groupings. The deviance residuals become quite large for the two groups with redlinks: Google and "other." The largest residuals appear to be associated with the observations for which a redlink was recorded, as evidenced by the lone large residual for Google. This makes sense, because recording a redlink is such an unusual event that most predicted probabilities of a redlink are extremely small, and therefore any observations of a redlink will be associated with great distance from those small predicted probabilities. This does not necessarily imply a poor fitting model, but rather is a reality of modeling unusual events. Overall the model appears to be acceptable but not excellent.

5.7.4 Model Parameter Interpretations

Table 5.13 shows parameter estimates and results of Wald z-tests for the logistic regression model of the probability of recording a redlink. Using "other" as the comparison group, the Google group is the only previous title classification that shows significance. Because all of the other previous title groups have 0 redlinks recorded, the logistic model struggles to assign a reasonable probability of a redlink to those groups and therefore there is a great deal of uncertainty in the estimated effects. Thus the large standard errors make it difficult to declare the effects significant. Nonetheless, all effects are negative, as is consistent

Table 5.13 *English Wikipedia Clickstream results of binary logistic regression of redlink.*

Predictor	Estimate	Std. error	z-value	p-value
Intercept	−5.815	0.011	−527.737	< 0.001
Bing	−16.399	116.143	−0.141	0.888
Google	−4.706	0.125	−37.795	< 0.001
Yahoo	−16.397	129.377	−0.127	0.899
Wikipedia	−16.235	79.024	−0.205	0.837
Main_Page	−16.375	190.561	−0.086	0.932
Empty	−16.332	52.667	−0.310	0.756

Residual deviance	114127 on 65460 d.f.
Null deviance	126834 on 65466 d.f.
AIC	114141

Table 5.14 *English Wikipedia Clickstream estimated odds ratios for redlink, with 95% confidence intervals, all odds compared to "other" group.*

Predictor	Odds ratio	95% Confidence interval
Bing	0.0000000755	(< 0.001, > 1000)
Google	0.00904	(0.007,0.015)
Yahoo	0.0000000757	(< 0.001, > 1000)
Wikipedia	0.0000000890	(< 0.001, > 1000)
Main_Page	0.0000000773	(< 0.001, > 1000)
Empty	0.0000000808	(< 0.001, > 1000)

with the descriptive evidence that the "other" group shows the largest number of redlink observations.

The negative effect estimate for Google suggests the expected likelihood of a redlink recording is lower for Google previous titles than for Other previous titles. Specifically, the odds of observing a redlink for Google is $\exp(-4.706) \approx 0.009$ times the odds of observing a redlink for Other previous titles, with associated 95% confidence interval (0.007,0.011). Negative effects in logistic regression models are often more easily interpreted in the reverse direction. The odds of observing a redlink for Other previous titles is $\exp(4.706) \approx 110.609$ times the same odds for Google, a relatively enormous increase. Odds ratio estimates and associated 95% confidence intervals for all previous title groups are shown in Table 5.14. The uncertainty associated with odds ratios for all groups other than Google is reflected in the incredibly wide confidence intervals produced.

5.7.5 Model Prediction

Predictions of the likelihood of observing a redlink can be made using the model prediction function,

$$\hat{\pi}_r = \frac{\exp(-5.815 - 16.399(\text{Bing}) - \cdots - 16.332(\text{empty}))}{1 + \exp(-5.815 - 16.399(\text{Bing}) - \cdots - 16.332(\text{empty}))}, \tag{5.16}$$

Table 5.15 *English Wikipedia Clickstream predicted probabilities of redlink for previous title groups.*

	Bing	Google	Yahoo	Wikipedia
Redlink probability	2.252e-10	0.0000270	2.258e-10	2.654e-10

	Main_Page	Empty	Other
Redlink probability	2.307e-10	2.409e-10	0.00297

where $\hat{\pi}_r$ indicates the predicted probability of a redlink. Predicted probabilities for each group are shown in Table 5.15.

We applied a logistic regression model to predict the probability of producing a redlink, which was effective even for the large number of observations involved in this analysis. Using the model it is clear that the "other" group shows the highest likelihood of a redlink, with Google the second most likely and all of the remaining previous title groups with near-zero chances of observing a redlink.

5.8 Summary

In this chapter we examined logistic regression models, which are used to model responses that are categorical in nature. We found that, instead of the typical means-based interpretations associated with normal linear regression models, logistic models involve making conclusions about probabilities and odds associated with outcome categories. We also found that these responses inherently have heterogeneous variation, and are usually related to predictors through a nonlinear function.

We modeled the probability of hypertension using a binary logistic regression model, allowing us to make conclusions about the probability of a participant developing hypertensive indicators, and about the odds of hypertension versus no hypertension. Using the logistic regression model, we found that the odds of hypertension increases for older individuals, for females, and for individuals with higher total cholesterol. The nonlinear logistic model allowed us to ensure all predictions of the probability of hypertension fell within the reasonable range of 0 to 1, unlike the predictions produced from the normal linear model applied in Chapter 3.

We modeled the probabilities associated with different levels of school bullying using an adjacent-categories ordinal multinomial logistic regression model. This model easily produced predicted probabilities associated with each level of bullying, and odds comparing each level to the immediately previous level. We found this model to be appropriate because the outcome of interest, bullying, was represented as five ordered categories. Using this model, we saw that the odds of greater levels of bullying increase for schools in areas of higher crime, for schools that offer student counseling, and for schools without uniform policies.

Using the very large number of observations in the English Wikipedia Clickstream data, we modeled the probability of encountering a redlink using a binary logistic regression model. This model allowed us to estimate the probabilities associated with a redlink for various previous page categories. We found that some of the diagnostic measures for logistic regression models become ineffective for data with large numbers of observations, but that the basic interpretations in terms of odds ratios and predicted probabilities remain the same. Using the logistic model, we found the "other" category to have the highest probability of resulting in a redlink, with "Google" showing a significantly smaller probability of leading to a redlink. Effects and predicted values showed an extremely high level of uncertainty because of the sparseness of the outcome of interest. In such situations with separation issues for a binary outcome, researchers have recommended alternative methods of estimation, such as penalized maximum likelihood.

5.9 Further Reading

A standard text for using logistic regression models is Hosmer et al. (2013). Hilbe (2015) provides an extremely accessible guide to using logistic models. Also easy to read is Kleinbaum and Klein (2010). Allison (2012) covers just about all aspects discussed in this chapter, with a focus on using SAS.

A comprehensive discussion of logistic and other categorical models is given in Agresti (1998), with a more accessible discussion available in Agresti (2007). A presentation of ordinal models is specifically given in Agresti (2010).

A very general treatment of logistic models is presented nicely in Dobson (2002) and very thoroughly in McCullagh and Nelder (1989).

Binary outcomes are treated specifically in Cox and Snell (1989) and also in Collett (2003).

A discussion of solutions to sparse binary outcomes or binary outcomes with separation can be found in Heinze and Schemper (2002) and in Allison (2012).

6

Count Response Models

6.1 Introduction

Counting in one form or another likely predates written human history. In modern society, counting is ubiquitous in many aspects of individual and collective human lives. One often-asked question of an individual is how many children are in his or her family. School systems are concerned with the related question of how many children are in a class or in an entire district, though the range of the counts in these cases is usually entirely different. A characteristic common to all counting systems is that counts are never less than zero, and some of these systems have maxima that may be treated as infinite: e.g., the number of grains of sand on all the beaches of the Earth, or the number stars in the observable universe. Some finite counting examples are the number of defects on an airplane, the number of radioactive decay products in a particle accelerator collision, or the number of students eligible for graduation term by term. Sports are teeming with examples of counting including the score from a round of golf, baseball's runs batted-in (RBI), and the number of goals averted in football. These examples demonstrate that count data permeate everyday human life.

The counting examples above suggest that count data have different types. The number of children in a nuclear family may be considered a tightly restricted range from zero to, say, twenty. However, the number of children in a school class is never zero. Batters on a baseball team may have a large number of players with no (zero) RBIs. The day-to-day sales of automobiles may range from zero and upwards based on a popular probability distribution such as the Poisson probability distribution or the negative binomial probability distribution. Count data statistical models rely on the probability distribution chosen for the parameter estimation method and reliability of the analysis outcomes.

Regardless of which probability distribution model is used to describe count data, one critical property of each model is that the counts are not transformed to approximate a normal distribution to allow the use of ordinary least squares estimation. Rather, the count distribution mean and variance are used to establish a nonlinear model link between the count response and the selected predictors. Normal linear regression, as seen in previous chapters, assumes that the response variable and the residual errors follow a normal probability distribution, have independent unexplained variance, and are linear in the model effects parameters. This allows model parameter estimation with ordinary least squares methods. Most count data violate one or more of these assumptions. While remedial measures such as transformations on the count variable or the predictors may be applied, these measures often do not satisfy the estimation method requirements, as demonstrated in Chapter 3. When the data have a mean-variance dependence, maximum likelihood estimation is an estimation method commonly employed.

The nonlinear count model form is

$$\mu_i = \exp(\beta_0 + \beta_1 x_{i1} + \cdots + \beta_p x_{ip}) \qquad (6.1)$$
$$= \exp(\beta_0)\exp(\beta_1 x_{i1})\cdots\exp(\beta_p x_{ip}),$$

which is known as a multiplicative model because each predictor does not additively affect the mean count. In Eq. (6.1), μ_i is the model-predicted mean, the x_{ij} are the observed values of explanatory variables (predictors), and β_j are the predictor coefficients, which must be estimated. Note that each natural exponential (anti-logarithm), exp(), term is multiplied.

Count data regression models assume the response variable follows a count-based probability distribution function as discussed in Chapter 2. Special cases of mean-variance relationships, too many zeros, not enough zeros, or counts completely absent require special consideration in count modeling.

6.2 Modeling Count Data

Count data comprise observations of nonnegative integer values. The distances between each level is a functional, discrete relationship among whole numbers without fractional interstices. Count responses are often transformed in an attempt to satisfy Gaussian parametric test assumptions rather than transforming the model predictor parameters. As was demonstrated in Chapter 3 and Chapter 4, such response transformations perform poorly, leading to bias in the estimated parameters. However, a class of regression models exists designed explicitly for count data.

A commonly used regression model for count data is Poisson regression. When count data are overdispersed, i.e., the mean and variance of the count data are not equal as with the Poisson model, remedial measures such as adjustments to the variance or the use of an alternate count distribution is employed. The negative binomial distribution is a popular alternate counts distribution for count regression. The negative binomial distribution has a functional relationship between the distribution mean and the variance, allowing for mean-variance relationship flexibility in count models. Situations leading to overdispersion include more zeros than expected, fewer zeros than expected, and, due to count clustering, overdispersion. These may be partitioned according to influences by individual predictors. An example of count clustering is standardized test outcomes in urban and rural schools. The higher density of students in urban schools can lead to wider variance in test scores than in lower student density rural schools. These various mean-variance conditions may be modeled by zero-inflated models, zero-deflated models, hurdle models, or heterogeneous negative binomial models. Table 6.1 compares several distribution models for the type of estimation method, type of response variable, allowance for zeros, variance structure, and management of overdispersion.

6.2.1 Poisson Models

Poisson regression is a probability-based count model. It assumes a discrete response described by a single parameter distribution. The model parameter is the mean, μ, which

Table 6.1 *Comparison of regression model characteristics.*

Distribution	Estimation method	Response variable	Allow zeros	Sample variance	Over-dispersion
Normal	OLS	Real	Yes	Homogeneous	No
Log-normal	OLS	Positive real	No	Homogeneous	No
Poisson	Likelihood	Nonnegative integer	Yes	Heterogeneous	No
Negative Binomial (NB)	Likelihood	Nonnegative integer	Yes	Heterogeneous	Yes
Heterogeneous NB	Likelihood	Nonnegative integer	Yes	Heterogeneous	Yes
Zero-inflated	Likelihood	Nonnegative integer	Yes	Heterogeneous	Yes
Zero-deflated	Likelihood	Nonnegative integer	Yes	Heterogeneous	Yes
Hurdle	Likelihood	Nonnegative integer	Yes	Heterogeneous	Yes

may be considered the expected number of times an event occurs within a fixed interval, often time, and thus may be considered as a rate. The single-parameter Poisson model assumes the mean and variance of the probability distribution are equal having what is known as equi-dispersion.

The mean and variance of the Poisson model are represented as

$$\text{Mean} = \mu, \qquad \text{Variance} = \mu. \tag{6.2}$$

Thus, a regression equation derived from Eq. (6.1) may be used for a count response, y, whose behavior may be influenced by a set of predictors with values anti-logged and added. The anti-log linear function of the predictors thereby keeps the count response variable on its original range.

Often, count data do not enjoy the Poisson assumption of equi-dispersion resulting in a Poisson dispersion statistic with a value greater than one, represented as

$$\text{Mean} = \mu, \qquad \text{Variance} = \mu + \phi\mu, \tag{6.3}$$

where ϕ is a nonnegative multiplier scaling the variance so it is larger than the mean. When $\phi \neq 0$, a Poisson regression model modification, known as a quasi-Poisson model, is one in which the standard errors of the predictor parameters are adjusted by an amount proportional to the value of ϕ. The values of the predictor parameters do not change from those if $\phi = 0$. The adjustment to the standard errors must be a linear function of ϕ, and the adjustment is reliable only for relatively small amounts of overdispersion when the excess dispersion is uniformly distributed on the response range.

6.2.2 Overdispersion

Overdispersion in Poisson models occurs when the value of the response variance is greater than the value of the response mean. It may arise when the response counts are clustered, or grouped contrary to what is expected from the distributional parameterization, thereby violating the likelihood requirement of the independence of observations. Overdispersion may cause the standard errors of the estimates to be deflated or underestimated, i.e., a variable may appear to be a significant predictor when it is in fact not. A check for overdispersion is performed with the dispersion statistic, $\mathcal{D}$,

$$\mathcal{D} = \frac{X^2}{N - N_p}, \tag{6.4}$$

where N is the number of observations and N_p is the number of parameters in the model. Then $N - N_p$ represents the unexplained variance (the residuals) degrees of freedom. For a Poisson model, the Pearson X^2 value is

$$X^2 = \sum_{i=1}^{N} \frac{(Y_i - \mu_i)^2}{\mu_i}, \tag{6.5}$$

where Y_i represents the observed counts, and μ_i is the mean and variance of Y_i. If overdispersion occurs, the variation in the data exceeds the expected variability based on the Poisson distribution resulting in $\mathcal{D}$ greater than 1. Small amounts of overdispersion have little effect on the model parameter estimates. Often, if $\mathcal{D} > 1.25$, a correction for overdispersion should be considered. See Hilbe (2014).

If overdispersion is present, then some corrective measures include adjusting the standard errors by scaling, applying sandwich or robust standard errors, or bootstrapping standard errors for the model. However, only the standard errors will be adjusted and not the regression coefficients, β. Situations exist, however, when the coefficients may be affected by overdispersion. When the coefficients are, or are suspected of being, influenced by overdispersion, an alternate count probability distribution may be utilized for count regression. The negative binomial distribution is often used as it contains additional parameters, called the dispersion or heterogeneity parameters, that may accommodate overdispersion.

6.2.3 Coefficient Interpretations

The statistics from fitting count models include estimates of the coefficients, the standard errors of the coefficients, a z-score determined from the coefficient value and its associated standard error, and a corresponding p-value.

Count-model coefficients interpretations allow for a variety of predictor types when the model output directly estimates the natural logarithm of the mean count. One way coefficient interpretation is affected is by whether a predictor is a continuous variable or a categorical variable. Continuous variable effects are described by ranges whereas categorical variable effects are described by level changes usually without metric relationships. For example, a binary categorical variable is either on or off, yes or no, interesting or not, for example. Multilevel categorical variables are similarly considered as nonmetric such as three geographic regions, small versus medium versus tall, or red or green or blue, as examples.

Consider a continuous variable predictor count model of the form $\mu = \exp(\beta_0 + \beta_1 x)$, x centered such that the mean of x is subtracted from each predictor observation. Suppose that the z-value and p-value each suggest the estimate, $\hat{\beta}_1$, is a significant influencer on the count response. When $x = 0$, $\exp(\hat{\beta}_0)$ is interpreted as the model mean at the mean value of x after inverting the centering transformation. A simple unit change in x gives the positive or negative percent change in the model mean as $\pm 100\%(1 - \exp(\hat{\beta}_1))$. For the percentage change in the model mean due to a specific deviation distance change in x, say $X_2 - x_1$, then $100\%(1 - \exp(\hat{\beta}_1(x_2 - x_1)))$ gives the desired change percentage.

For a binary predictor x that takes either of two levels, say 0 or 1, the count model behaves as does the continuous variable model with $\mu = \exp(\beta_0 + \beta_1 x)$. The interpretation differs from the continuous predictor model in that the only change in the model mean is when $x = 1$. Hence, $100\%(1 - \exp(\hat{\beta}_1))$ is the percent change in the count mean when $x = 1$ over the counts mean of $\exp(\beta_0)$ when $x = 0$.

When x is a multilevel categorical predictor variable, the levels of x are expanded as $x_1 = 1$ when $x = 1$, $x_2 = 1$ when $x = 2$ and $x_3 = 1$ when $x = 3$, and zero otherwise for a three-level categorical variable. The subscripts on the indicator variables are extended to account for more than three levels. The count model becomes $\mu = \exp(\beta_0 + \beta_2 x_2 + \beta_3 x_3)$, using the first level of $x = 1$ as the reference level. Thus the variable x_1 and its associated parameter β_1 are not used. Some statistics packages use the highest level of either a binary or a multilevel categorical variable as the reference. If this is the case, then the three-level categorical model is written using $\beta_1 x_1 + \beta_2 x_2$, leaving out $\beta_3 x_3$. If $x = 1$, then $x_2 = 0$ ad $x_3 = 0$, leaving the model mean as $\exp(\hat{\beta}_0)$, just as with the continuous and binary predictor models. If $x = 2$, then the percentage change estimate in the model mean is $100\%(1 - \exp(\hat{\beta}_2))$, which is the percentage change in the model mean when changing from the categorical predictor variable level 1 to level 2. The calculation for determining the percent change in the counts mean for $x = 3$ is similar, but β_2 is replaced with $\hat{\beta}_3$. The percent change in the model mean for a change from $x = 2$ to $x = 3$ is $100\%(1 - \exp(\hat{\beta}_3 - \hat{\beta}_2))$.

To model counts with both a centered continuous predictor and a binary predictor, the model is

$$\mu = \exp(\beta_0 + \beta_1 x_1 + \beta_2 x_2 + \beta_{12} x_1 x_2), \tag{6.6}$$

in which x_1 is the binary predictor variable and x_2 is the centered continuous predictor variable. The interaction is included as $x_1 x_2$. Centering the continuous predictor variable simplifies the interpretation of this model, particularly the interaction term: there is no continuous variable parameter estimate that is added to $\hat{\beta}_0$ when $x_1 = 0$ and the continuous predictor is set to its mean value forcing $x_2 = 0$. When $x_1 = 0$, the various ranges of x_2 may be examined without the interaction term similar to the continuous predictor example above. When $x_1 = 1$ and the continuous predictor is set to its mean value, the analysis is of a single binary predictor model, again as shown previously. The case of interest in this simple interaction model is when $x_1 = 1$ and the continuous predictor is not held to its mean value. First assume a unit change in the continuous variable x_2. Then the percent change in the model mean is $100\%(1 - \exp(\hat{\beta}_1 + \hat{\beta}_2 + \hat{\beta}_{12}))$. For a change in the continuous predictor other than a unit change, the percentage change in the model mean is $100\%(1 - \exp(\hat{\beta}_1 + \hat{\beta}_2(x_b - x_a) + \hat{\beta}_{12}(x_b - x_a)))$.

Any combination of continuous and categorical variables is analyzed similarly to the above cases. Note that in each case of the continuous and categorical count models, β_0 cancels out when determining the percentage change in the model mean using the predictor estimated parameters.

The interpretation of each coefficient in the quasi-Poisson model and the negative binomial model is the same as that of the Poisson model. The differences among these models are the values of the standard errors, and hence, the amount of variability in the response that is accounted for by the predictors. Different values of the standard errors may change the values of the z-values and the associated p-values, thus changing the level of significance of the predictors. The quasi-Poisson model estimation scales the standard errors by the value of the dispersion parameter, thus adjusting these errors to a level that is expected if no overdispersion is present. The adjustment effectively inflates the values of these errors which reduces the significance of each predictor's contribution to the variability of the counts. The negative binomial model has a second parameter to accommodate overdispersion, and hence the standard errors are adjusted accordingly.

6.2.4 Negative Binomial Models

The negative binomial (NB) distribution is traditionally represented as the probability of observing y failures before the rth success in a series of Bernoulli trials. It can also be formulated as a Poisson model with a gamma-distributed variance. The NB model, as a Poisson-gamma mixture model, denoted as NB2, is appropriate to use when the overdispersion in an otherwise Poisson model takes the shape of a gamma distribution.

The NB distribution function has two parameters, μ and k, allowing more options for variance modeling than the single parameter Poisson distribution. The mean and variance may be represented as

$$\text{Mean} = \mu; \qquad \text{Variance} = \mu + \frac{\mu^2}{k} = \mu + \alpha\mu^2. \qquad (6.7)$$

The NB distribution has assumptions similar to the Poisson distribution with the exception that it has a dispersion parameter, $\alpha = 1/k$, to accommodate more varieties of count distribution shapes than does the Poisson model. As the dispersion parameter, α, approaches 0, the NB variance equals the mean which degenerates to the single-parameter Poisson distribution.

Note that if count data have different clusters each with a possibly different gamma-shaped dispersion indicating nonconstant correlation within the data, and if the NB Pearson χ^2 dispersion statistic is greater than one, then the NB model is likely overdispersed; i.e., the count data are likely overdispersed regardless of the probability distribution model used. Random effects and mixed-effects count models may be appropriate to account for clustered dispersion cases, as discussed in Chapter 8.

If the dispersion statistic is less than one, then the count data are said to be underdispersed. The NB model is not appropriate for managing underdispersion; however, the Poisson and quasi-Poisson models are. We do not discuss underdispersed count data.

The heterogeneous negative binomial (NB-H) model is used to identify each predictor's contribution to over-dispersed count data. The NB-H model partitions the single NB2 dispersion parameter into values identified for each predictor. If the p-value for any predictor suggests the predictor accounts for a significant proportion of the response variability, then this predictor may be thought of as significantly influencing the level of dispersion. The

NB-H model may be used not just for count clustering diagnosis, but also as an analytic, predictive count model.

6.2.5 Zero-Inflated Models

The definitions of both the Poisson and NB models mandate an expected number of zeros in the data depending on each distribution's mean and variance parameters. If the count data have more than the expected number of zeros, a zero-inflated model, also known as an excess-zero model, should be considered, i.e., either a zero-inflated Poisson model or a zero-inflated NB model. There are zero-inflated versions of, e.g., inverse Gaussian count models as well, but this discussion is restricted to the zero-inflated NB model.

Data sets with more (or fewer) zeros than expected from a specified count distribution may be thought of as having the zeros generated by a process different than that which generates the nonzero counts. For counts with an inflated number of zeros, the zeros and counts greater than zero may be modeled from two overlapping processes: a binary process to represent the presence of zeros, and a count process based on a NB distribution to represent the distribution's expected number of zeros along with the nonzero counts. The overlap is from both processes generating zero counts. The binary process is modeled using logistic regression, while the NB process is a regression model. These two overlapping processes to model count data is called a zero-inflated model. Because the processes overlap modeling zeros, zero-inflated models include the assumption that observed zeros can occur from either process. For example, in a model of school attendance, students may have zero absences because of good health (as part of the NB count process), but individuals may have no absences but are still not in class because of sanctioned extracurricular activities (as part of the binary process). The two components of an excess-zero model can be written as

$$\pi_i = \frac{\exp(\gamma_0 + \gamma_1 x_{i1} + \cdots + \gamma_p x_{ip})}{1 + \exp(\gamma_0 + \gamma_1 x_{i1} + \cdots + \gamma_p x_{ip})},$$

$$\mu_i = \exp(\beta_0 + \beta_1 x_{i1} + \cdots + \beta_p x_{ip}). \tag{6.8}$$

In Eq. (6.8) the x_{ij} represent observed values of predictors, which need not be the same between the two components of the model, while the γ_j and β_j represent the unknown coefficients for the logistic and count components of the model, respectively, which are used to estimate π_i and μ_i.

6.2.6 Zero-Deflated Models

A zero-deflated count model has the expected number of zeros less than that of the selected count distribution such as the Poisson distribution or the NB distribution. As with the zero-inflated count models, the zeros may be considered to be generated by a process different from the process that generates the expected number of zeros and nonzero counts generated by the selected distribution. As with the zero-inflated model in Eq. (6.8), a count distribution such as the NB distribution is used model the expected numbers of zeros and the nonzero counts, while a logistic model is used to determine the deflated zero probability.

6.2.7 Hurdle Models

An alternate model for counts with inflated or deflated numbers of zeros is to model the inflated zero counts, and the expected number of zeros and counts greater than zero as two separate process rather than a mixture of these two processes as with the zero-inflated or zero-deflated models. The hurdle model construction first partitions the zero counts from the nonzero counts. A binary process is used for the zero counts where a value of 1 represents a nonzero count, and a value of 0 represents a zero count. The binary process then is modeled with a logistic model just as with the zero-inflated model. Next, the hurdle construction "hurdles" the zeros and models the nonzero counts with a zero-truncated Poisson or NB model, or some other count probability distribution. The difference between the hurdle model and the zero-inflated model is that the, say, NB model excludes the counts of zero rather than treating the zeros as a mixture as in the zero-inflated model. Hence, the zero and nonzero counts are completely segregated. Also, the hurdle threshold may be a count higher than zero, e.g., all counts greater than 1.

The hurdle model can be written similarly to the zero-inflated model given in Eq. (6.8), but estimation follows a likelihood process that acknowledges the separate zero and nonzero processes. Interpretations of parameters in the hurdle model must be made with greater care, as the coefficients in the count component of the hurdle model represent conditional effects, conditioned on clearing the hurdle. That is, the β_j in the hurdle version of Eq. (6.8) represent effects of predictors for the population of individuals who clear the hurdle, in this case who have positive counts.

6.3 Fire-Climate Analysis: Decade Counts

Fire-Climate Interactions in the American West since 1130 data (Trouet et al., 2010), archived by World Data Center for Paleoclimatology, Boulder, Colorado, and the National Oceanic and Atmospheric Administration (NOAA) Paleoclimatology Program (www.ncdc. noaa.gov/paleo/impd/paleofire.html), consist of annually counted, tree rings as fire records for four regions in the American West that extend back to 1130. Chapter 1 introduces these data in more detail, and here we are concerned with the American West regions Pacific Northwest (PNW), Northern California (NC), and Interior West (IW), spanning the years from 1130 to 2004.

For each of the three regions (PNW, NC, and IW), fire-scar records were selected from individual trees that were exposed to fire as indicated in the tree rings. In each year, the number of trees are counted, and the count aggregated into decadal counts. It is these fire-indicating ring counts we will model by region. Each count is the number of trees in a decade and region that indicates the presence or absence of fire. We use the following variables:

- Decade: the decade, from 1130 through 2000
- Region: the region (Pacific Northwest, Interior West, and Northern California)

6.3.1 Exploratory Data Analysis

Recall from Chapter 1 the distribution of the decadal fire counts is not a single parameter Poisson distribution (Figure 6.1). This situation is an opportunity to explore several count model options.

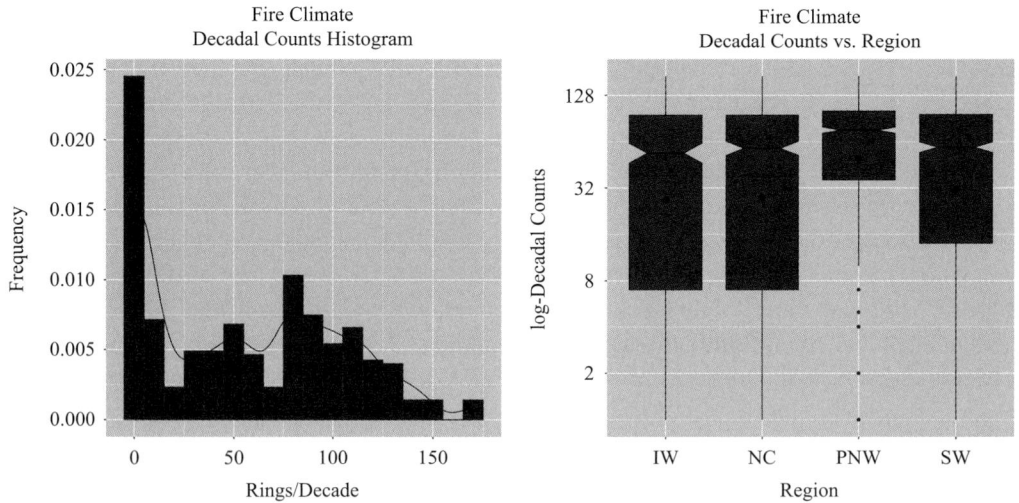

Figure 6.1 The left-hand panel is a histogram of the number of fires indicated by tree ring scars by decade. The black solid curve is a smoothed density fit to the histogram. The right-hand panel is a box-whisker plot of the log of fire counts by region. Each of the box-whisker plots suggest the distributions are non-Gaussian.

A critical test of these data is for overdispersion; i.e., is the mean ($\bar{y}$) of these count data equal to the variance (Var(y)) of these data. Recall from the exploratory data analysis that Var(y) = 451.8568 and $\bar{y}$ = 25.63868, so clearly, as Var(y) $\gg \bar{y}$, there is evidence of overdispersion.

Another critical test of the fire counts is for inflated numbers of zeros over what is expected from a selected counts distribution. The histogram in Figure 6.1 shows an inflated number of zeros for the Poisson distribution, and we should test this is actually the case. The number of zeros in the fire data is 357. The expected number of zeros from a Poisson distribution with mean $\hat{\mu} = 25.639$ as calculated from the fire data is 7.333^{-12}, which is practically zero. Clearly, the actual number of $357 \gg 0$ indicates the fire data are zero-inflated.

6.3.2 Poisson Model

We tested the fire data and showed these count data are overdispersed. Therefore rather than beginning with a traditional single parameter Poisson count model, we shall use scaling to adjust the standard errors to emulate a dispersion statistic of unity. We know this model is not likely to be adequate, but it will give a basis for comparing goodness-of-fit to other count models we'll discuss below. The response is the number of tree rings indicating a presence or absence of fire in the given decade, and the predictor variables are the region and the decade. (Note that we assume decades are uncorrelated.) The model is

$$\mu = \exp\left(\beta_0 + \beta_2(\text{NC}) + \beta_3(\text{PNW}) + \beta_4(\text{decade})\right), \tag{6.9}$$

Table 6.2 *Fire-Climate data Poisson model goodness-of-fit statistics.*

GOF measure	Estimate
Deviance statistic	1 542.463
Deviance statistic X^2	7.815
p-value	< 0.001
Dispersion parameter	16.327
AIC	74 676.847

Table 6.3 *Fire-Climate data quasi-Poisson model goodness-of-fit statistics.*

GOF measure	Estimate
Deviance statistic	1 542.463
Deviance statistic X^2	7.815
p-value	< 0.001
Dispersion parameter	16.327

where the μ is the model count mean, NC is the North Central region, PNW is the Pacific Northwest region, the Interior West region is the reference level, decade is the decade after mean-centering, and the $\boldsymbol{\beta}$ is the vector of parameters to be estimated.

The intercept, β_0, is the linear predictor of μ when the regions and mean-centered decade are set to zero. The parameters β_2 and β_3 are the height changes relative to the intercept, and the steepness of the respective slopes is an indication of the amount of change in the response due to each predictor.

Poisson Model Adequacy

The coefficients analysis of the Poisson model is of little consequence if the model is not adequate to represent the data. Table 6.2 gives two model adequacy measures: deviance and AIC.

The deviance statistics indicate the Poisson model significantly (p-value $\ll 0.05$) deviates from the saturated model (the model that is overparameterized such that it essentially reproduces the observed data). Further, the dispersion parameter is $16.327 \gg 1$, indicating the variance is over 16 times that of the mean. These two statistics indicate the Poisson model is a poor fit to the fire data.

The deviance statistics (Table 6.3) for the quasi-Poisson model are the same as for the Poisson model. The difference is the dispersion parameter which, for the Poisson model, is assumed to be unity. The quasi-Poisson model, however, estimates a dispersion parameter that is used to adjust the standard errors. A likelihood function is not used to estimate the parameters of the quasi-Poisson model and hence an AIC statistic is not available.

We shall use AIC as one of the statistics to compare the Poisson, NB, zero-inflated NB (ZINB), and hurdle models described below to assess whether any of these is superior to the Poisson and quasi-Poisson models.

Table 6.4 *Fire-Climate data NB2 model goodness-of-fit statistics.*

GOF measure	Estimate
Deviance statistic	126.365
Deviance statistic X^2	7.815
p-value	0.000
Dispersion parameter	1.106
AIC	27 306.471

6.3.3 Negative Binomial Models

A benefit of using the negative binomial distribution for modeling counts over the use of the Poisson distribution is that the model may better suit the variance-to-mean relationship. The NB two-parameter probability distribution structure is the feature that permits tailoring to the mean-variance structure. Recall that the Poisson distribution assumes data have a mean equal to its variance, a restriction that rarely occurs in count data. The NB distribution, however, has at least 22 ways of managing overdispersion via the mean-variance relationship. A common incarnation of the NB model is the Poisson-gamma mixture model (NB2).

NB Model Adequacy

The deviance statistic (127.1) in Table 6.4 is much larger than the χ^2 value of 7.815 for three degrees of freedom, so we find the fire NB2 model is not adequate, which is not surprising because of the large overdispersion in the fire data.

The dispersion parameter (1.106) in Table 6.4 is approximately unity and suggests the mean and variance are approximately equal. We see that the NB2 dual parameter estimates of the mean and variance has accounted for the fire data overdispersion much better than either the Poisson or quasi-Poisson models. Even though the NB2 goodness-of-fit shows the fire NB2 model is not an equivalent representation of the fire data to the saturated model, the NB2 manages the overdispersion quite well.

The NB model has an AIC of 27 317 which is much less than the AIC of the Poisson model's AIC of 74 521. We may conclude that the NB2 model outcomes are more reliable than those of the Poisson model.

Often, the heterogeneous negative binomial (NB-H) model identifies which predictors may lead to a poor fit to the count data. Specifically, which predictors may contribute to clustered counts and hence heterogeneity of variance. As the NB2 fit is marginal, we turn to the NB-H model to see if remedial measures may be identified.

The heterogeneous NB (NB-H) model is used to identify each predictor's contribution to overdispersion. The NB-H model decomposes the single NB2 dispersion parameter (α) into values identified with each predictor. If the p-value is smaller than the selected threshold value for any predictor, then these predictors may be thought of as significantly influencing the level of dispersion. While we have no concern that the NB2 overdispersion is an issue, we explore the use of the NB-H to demonstrate its diagnostic capability.

Table 6.5 *Fire-Climate count of fires by region and decade NB model parameter estimates with robust lower and upper confidence limits (LCL and UCL).*

Predictor	Estimate	Std. error	z-value	p-value	LCL	UCL
Negative binomial model estimates						
Intercept	3.906	2.32×10^{-2}	168.193	< 0.001	3.860	3.951
$\hat{\beta}_2$ (NC)	-0.040	2.34×10^{-2}	-1.697	0.090	-0.086	0.006
$\hat{\beta}_3$ (PNW)	0.180	2.68×10^{-2}	6.706	2×10^{-11}	0.127	0.233
$\hat{\beta}_4$ (decade)	0.007	9.59×10^{-5}	76.510	< 0.001	0.007	0.008
Source of heterogeneity estimates						
Intercept	1.26861	5.24×10^{-2}	-24.199	< 0.001	-1.37136	-1.166
$\hat{\beta}_{2p}$ (NC)	-0.01392	7.23×10^{-2}	-0.193	0.847	-0.15562	0.128
$\hat{\beta}_{3p}$ (PNW)	-0.02472	7.02×10^{-2}	-0.352	0.725	-0.16236	0.113
$\hat{\beta}_{4p}$ (decade)	0.00914	2.03×10^{-4}	45.091	< 0.001	0.00875	0.010
	Dispersion parameter	1.384				
	AIC	25 737.42				

The estimated predictor parameters in Table 6.5, $\beta_{ip}, i = 0, 2, 3, 4$, are the parameters used to assess the predictor influence on dispersion. We see that the intercept (β_{0p}), which, for all other predictor values held at zero represents the IW response to the count of tree rings indicating fires, strongly influences the dispersion. The same may be said of the decade predictor (β_{4p}). As such, we may try a NB2 model with the decade predictor removed. A level-reduced NB2 model is suggested by the AIC (25 737) being the lower AIC of the Poisson and NB models. However, we already know the fire data have too many zeros, and hence, we suggest constructing a zero-inflated NB model to account for potentially unstable dispersion.

The parameters for the NB-H model can be analyzed as we did for the NB2 model parameters, in some cases, the NB-H model may be the most appropriate model. However, we see that the NB-H model dispersion statistic of 1.384135 is greater than the dispersion statistic of 1.106 of the NB2 model which suggests the NB2 model better manages the fire data overdispersion. However, the NB2 AIC = 27 317 is larger than the HB-H AIC of 25 737 thus evincing a case for preferring the NB-H model. We examine the outcome of the zero-inflated negative binomial model prior to settling on either the NB2 or the NB-H model in the event accounting for the large number of zeros will provide a superior model of the fire data.

6.3.4 Zero-Inflated NB Models

The definitions of both the Poisson and NB models mandate an expected number of zeros in the data depending on each distribution's parameter values. We showed earlier that the fire data have more than the expected number of zeros, and hence we should consider a zero-inflated NB (ZINB) model. There are zero-inflated versions of, e.g., Poisson and inverse Gaussian models as well, but we restrict this discussion to the ZINB. See the references for in-depth discussions of the various forms of the zero-inflated models.

Table 6.6 *Fire-Climate ZINB assessment statistics.*

GOF measure	Estimate
Log-likelihood	-1.336×10^4 on 9 df
Dispersion parameter	0.598

Table 6.7 *Fire-Climate ZINB Vuong nonnested hypothesis test statistic (test statistic is asymptotically distributed N(0,1) under the null that the models are indistinguishable).*

Test	Vuong z-value	H_A	p-value	Use model
Raw	11.187	ZINB > NB2	$\ll 0.001$	ZINB
AIC-corrected	11.105	ZINB > NB2	$\ll 0.001$	ZINB

Let us assume that a zero in the tree ring data represents a ring sample for which there are no fires rather than the zero representing a nonsampled ring. That is, all the zeros represent the absence of fire for the designated ring and hence for the represented decade. Then we may model the binary process with a logistic model (see Chapter 5) in which a value of 1 indicates a zero, and a value of 0 represents a nonzero count.

ZINB Model Adequacy

The dispersion parameter of 0.598 in Table 6.6 is less than one with the variance 0.598 times less than the mean, and certainly less than the dispersion statistic of 1.106 of the NB2 model. This underdispersion calls into question whether the ZINB model better accounts for the overdispersion in the fire data than does the NB2 model. Rather than use the AIC statistic for further comparing the ZINB and NB2 models, we use a comparison test proposed by Vuong (Vuong, 1989). The reason is that the ZINB and NB2 models are not considered nested models, i.e., two different modeling processes were used on the ZINB model and the NB2 model to produce accountability statistics and parameter estimates.

Table 6.7 provides two measures for comparing the ZINB model with the NB2 model. The assumptions to satisfy Vuong's nonnested hypothesis test statistic are stringent, and care in its use is in order. The table gives Vuong's z-value for the uncorrected test (the "Raw" row). The uncorrected test does not account for the additional parameters resulting from the logistic model portion of the ZINB model. As we have focused on the AIC statistic for model comparison, so shall we continue with the AIC-corrected row of the table (though for our ZINB and NB2 models, each row's statistic gives the same result). The AIC-corrected statistics (large positive z-value) suggest that the ZINB model is a closer representation of the underlying data processes than is the NB2 model. This is confirmed by comparing the NB2 AIC = 27 317, which is larger than the ZINB AIC = 26 229.

While Vuong's test indicates the ZINB model is superior to the NB2 model, we still must reconcile the dispersion statistic disparity between these two models. A possible solution is to use a hurdle model to describe the excess zeros, and possibly account for the underdispersion of the ZINB.

Table 6.8 *Fire-Climate ZINB NB model coefficients.*

Predictor	Estimate	Std. error	z-value	p-value
Negative binomial model estimates				
Intercept	3.261	0.021	156.509	$\ll 0.001$
$\hat{\beta}_2$ (NC)	0.008	0.036	0.222	0.825
$\hat{\beta}_3$ (PNW)	0.142	0.038	3.730	< 0.001
$\hat{\beta}_4$ (decade)	0.002	< 0.001	19.058	$\ll 0.001$
Predictor	Estimate	se	z-value	p-value
Zero logit model estimates				
Intercept	-1.956	0.077	-25.330	$\ll 0.001$
$\hat{\beta}_2$ (NC)	-0.263	0.144	-1.826	0.068
$\hat{\beta}_3$ (PNW)	-0.775	0.188	-4.119	$\ll 0.001$
$\hat{\beta}_4$ (decade)	-0.001	< 0.001	-3.242	0.001

ZINB Model Parameter Estimates

Table 6.8 displays the statistics from fitting the number of tree rings indicating fires by region and by decade (mean-centered). The model intercept value ($\hat{\beta}_0 = 3.261$) has a very large z-value (156.509) and the p-value is less than any common threshold value, suggesting this intercept is not zero; i.e., the model mean value is 3.261 when decade is zero (the mean value of decade after the centering transformation) in the IW region. The mean of decade is 1591.7, thus making the mean decade between 1590 and 1600. The expected count for this decade in the IW region then is, using the anti-log $e^{3.261} = 26.0646$, so the expected count is 26.

The coefficient of the decade predictor (0.002) has a large z-value with p-value much less than common thresholds. This suggests centered decade accounts for a significant amount of the variability in the response. The change in the response due to a unit change in decade (i.e., a ten-year change) is an increase of 0.002. This corresponds to a change in the mean count of $e^{0.002} = 1.002$, which is an approximate 0.2% increase of tree rings indicating fires per decade in the IW region. Essentially, in the IW region over the span of tree ring sampling time, there is a 2% increase in the number of fires in each century following the century encompassing the mean decade.

We now repeat the IW region analysis for the NC region. When we hold the centered decade at zero, the response change of the NC region over the IW region is 0.008, which corresponds to an increase in tree ring counts of $e^{0.008} \approx 0.80\%$, suggesting the NC region has approximately four times more fires than does IW in the decade of 1590 to 1600. The change in the response due to a unit change in decade for the NC region is $0.008 + 0.002 \approx 0.010$ which is a $e^{0.01} \approx 1.0\%$ increase in tree ring counts per decade advancing (or retreating) in time from the mean decade.

The process for determining the fire rate in the PNW region over the IW region is as we showed for the NC region. To compare the PNW region with the NC region, first hold the decade change at zero, then difference the NC coefficient from the PNW coefficient to obtain the change in the response. Thus, $0.142 - 0.008 = 0.134$ is a response increase for the

Table 6.9   *Fire-Climate hurdle assessment statistics.*

GOF measure	Estimate
Log-likelihood	< -0.001 on 9 df
Dispersion statistic	0.616

Table 6.10   *Fire-Climate hurdle model Vuong nonnested hypothesis test statistic (test statistic is asymptotically distributed N(0,1) under the null that the models are indistinguishable).*

Test	Vuong z-value	H_A	p-value	Use model
Raw	−9.2303	hurdle > ZINB	≪ 0.001	ZINB
AIC-corrected	−9.230	hurdle > ZINB	≪ 0.001	ZINB

PNW region over the NC region. This gives an $\approx 14.34\%$ increase in fire rate for PNW over NC. When the coefficient of decade is added to the difference of the coefficients of PNW and NC, the decadal rate of PNW over NC results.

The logistic model in the ZINB essentially gives the odds of any of the predictor variables being zero. The odds that decade produces a count of zero is $e^{-0.0009144} = 0.9990860$. The odds of 1 stands to reason as the decade data are centered. Given a zero for the centered decade, the odds of the IW region producing a count of zero is $e^{-1.956} = 0.1414$, which suggests a IW not having a zero count for the mean decade is approximately 7 to 1. The odds of no fires in the PNW region is $e^{-0.775} = 0.4607$, thus the odds of a fire in the mean decade are approximately 2 to 1.

The ZINB model shows itself to be a viable model for the fire data, and is decidedly superior to the normal linear model in Chapter 3. This is due in large part to accounting for the excess zeros in the fire data. An alternate model for counts with an inflated number of zeros is the hurdle model, and we investigate this model to be thorough.

Table 6.9 gives the values of the log-likelihood and the dispersion statistic (0.616) which is less than the ZINB model dispersion statistic (0.740). Table 6.10 provides two Vuong measures for comparing the hurdle NB2 model with the ZINB model. The AIC-corrected statistics (large negative value of the z-statistic) suggest that the ZINB model is a closer representation of the underlying data processes than is the NB2 model. Based on these statistics, we disregard the hurdle model in favor of the ZINB model.

We may be able to achieve an adequate hurdle model by setting the threshold to a higher count, e.g., hurdle values less than two or perhaps three.

The analysis of the Fire-Climate data allowed us to try a Poisson, quasi-Poisson, two versions of the negative binomial, and two types of excess zero models. The EDA clearly showed the fire data have excess zeros. We showed that the zero-inflated negative binomial had the most desirable set of fit statistics for the models constructed. A major point to this analysis is the iterative nature of model construction and selection. In the case of the fire data as revealed by the EDA, we ordinarily would have begun model construction and assessment with excess zero models. Fortunately, there are many choices for counts models making finding the one that best matches the data and study intention a very likely circumstance.

6.4 SSOCS Analysis: Annual Suspensions

We briefly revisit the analysis of Chapter 4, Section 4.5, in which we modeled the number of annual suspensions using various school characteristics. In Chapter 4, we applied a combination of outcome transformations and weighted least squares analyses to try to satisfy the normal linear model assumptions. In this section we will present a more appropriate excess-zero model, along with the additional interpretations such a model allows.

Exploratory analyses were presented in Chapter 1 and also in Chapter 4. Recall that the outcome of interest, annual suspensions, was seen to be highly right-skewed. As a count response, this is to be expected. We can also see that 1 918 of the 2 560 responses are zeros, or 74.922% of the count responses are zeros. This is clearly an excess of zeros according to what would be expected under a Poisson or a negative binomial distribution. Therefore we will apply one of the excess-zero models to account for this property.

For this application, it is reasonable to expect that the sub-population of schools that report zero suspensions in an academic year is distinct from the sub-population of schools that report suspensions. In other words, we believe schools can be split into two groups: schools that tend to suspend students and schools that tend not to suspend students. Therefore we will apply a hurdle regression model, which imposes the assumption that the schools reporting suspensions are distinct from the schools reporting no suspensions. We will use the negative binomial version of the hurdle model to accommodate possible overdispersion.

6.4.1 Hurdle Negative Binomial Model

An appropriate hurdle negative binomial model can be written in two components. The logistic component of the model will be used to model the likelihood of observing a positive number of suspensions,

$$
\begin{aligned}
\ln\left(\frac{\pi_s}{1-\pi_s}\right) = {} & \gamma_0 + \gamma_1(\text{uniforms}) + \gamma_2(\text{metal detectors}) + \gamma_3(\text{tipline}) \\
& + \gamma_4(\text{counseling}) + \gamma_5(\text{moderate crime}) + \gamma_6(\text{high crime}) \\
& + \gamma_7(\text{discipline training}) + \gamma_8(\text{behavioral training}) \\
& + \gamma_9(\text{insubordination}) + \gamma_{10}(\text{limited English}) \\
& + \gamma_{11}(\text{discipline} \times \text{moderate crime}) + \gamma_{12}(\text{discipline} \times \text{high crime}) \\
& + \gamma_{13}(\text{behavioral training} \times \text{moderate crime}) \\
& + \gamma_{14}(\text{behavioral training} \times \text{high crime}),
\end{aligned}
\tag{6.10}
$$

where π_s represents the probability of reporting a positive number of suspensions. The count component of the model, conditional on observing positive counts, will be a zero-truncated negative binomial regression only for those schools reporting more than 0 suspensions,

$$
\begin{aligned}
\mu_s = {} & \beta_0 + \beta_1(\text{uniforms}) + \beta_2(\text{metal detectors}) + \beta_3(\text{tipline}) \\
& + \beta_4(\text{counseling}) + \beta_5(\text{moderate crime}) + \beta_6(\text{high crime}) \\
& + \beta_7(\text{discipline training}) + \beta_8(\text{behavioral training}) \\
& + \beta_9(\text{insubordination}) + \beta_{10}(\text{limited English}) \\
& + \beta_{11}(\text{discipline} \times \text{moderate crime}) + \beta_{12}(\text{discipline} \times \text{high crime}) \\
& + \beta_{13}(\text{behavioral training} \times \text{moderate crime}) \\
& + \beta_{14}(\text{behavioral training} \times \text{high crime}),
\end{aligned}
\tag{6.11}
$$

where μ_s represents the mean number of suspensions among schools that report positive numbers of suspensions. For this model we have assumed the same predictors for both the logistic and count components of the model, although that is not a necessity of a hurdle model. The γ_j will provide information about effects on the likelihood of reporting any suspensions, while the β_j will describe the expected effects on the number of suspensions for schools that report suspensions.

6.4.2 Model Fit

Residuals from the negative binomial hurdle regression model range from -0.431 to 13.000. The large positive residuals indicate that there are extremely large observed values whose magnitudes are not completely captured by the model. Compare to the range of -175.000 to 7.569 for the model selected in Chapter 4, using a log-transformed response and the inverse of the number of insubordinate students as weights in the estimation. It is clear that the hurdle model has superior fit just based on the residuals. More importantly, we *know* that the hurdle model is more appropriate than a weighted normal regression model, because the response is a count with almost 75% of the observations as zeros.

6.4.3 Model Interpretations

Estimates of effects and tests of significance for the logistic component of the hurdle model are provided in Table 6.11.

Results of the Wald tests show that the availability of a tipline to report issues, and the number of insubordinate students are significantly associated with a greater likelihood of reporting suspensions. Specifically, when compared to schools without a tipline, the odds of reporting suspensions versus reporting zero suspensions increases by a factor of $e^{0.310} \approx 1.363$ for schools with a tipline. Test results also show a significant increase in the likelihood of reporting suspensions for schools in areas of moderate crime, as compared to schools in areas with low crime. Specifically, the odds of reporting suspensions increases by a factor of $e^{0.842} \approx 2.321$ for schools in areas of moderate crime. Surprisingly, the same effect was not seen for schools in areas of high crime. It may also be of interest that the training types in discipline policies and in positive behavioral interventions are not associated with any meaningful change in the likelihood of reporting suspensions.

Table 6.12 shows the effects estimates and Wald tests for the count component of the hurdle model.

Table 6.11 *School Survey on Crime and Safety results of logistic component of negative binomial hurdle regression of annual suspensions.*

Predictor	Estimate	Std. error	*t*-value	$Pr(>|t|)$
Intercept	−1.826	0.232	−7.884	< 0.001
Uniforms	0.255	0.136	1.876	0.061
Metal detectors	−0.244	0.298	−0.817	0.414
Tipline	0.310	0.097	3.194	0.001
Counseling	0.267	0.217	1.233	0.218
Moderate crime	0.842	0.286	2.940	0.003
High crime	0.378	0.579	0.652	0.514
Discipline training	0.178	0.132	1.350	0.177
Behavioral training	−0.011	0.138	−0.076	0.939
Insubordinates	0.002	0.0002	6.412	< 0.001
Percent age limited English	−0.007	0.004	−1.908	0.056
Discipline × moderate crime	0.131	0.285	0.461	0.645
Discipline × high crime	−0.121	0.473	−0.256	0.798
Behavioral × moderate crime	−0.461	0.310	−1.487	0.137
Behavioral × high crime	−0.084	0.602	−0.139	0.890

Table 6.12 *School Survey on Crime and Safety results of count component of negative binomial hurdle regression of annual suspensions.*

Predictor	Estimate	Std. error	*t*-value	$Pr(>|t|)$
Intercept	1.397	0.453	3.084	< 0.001
Uniforms	0.093	0.208	0.446	0.655
Metal detectors	0.650	0.494	1.316	0.188
Tipline	0.179	0.156	1.147	0.251
Counseling	−0.209	0.357	−0.587	0.557
Moderate crime	0.617	0.435	1.419	0.156
High crime	0.821	1.005	0.816	0.414
Discipline training	0.063	0.220	0.287	0.774
Behavioral training	0.089	0.229	0.387	0.698
Insubordinates	0.004	0.001	5.544	< 0.001
Percent age limited English	−0.003	0.006	−0.464	0.643
Discipline × moderate crime	−0.243	0.457	−0.531	0.595
Discipline × high crime	0.293	0.737	0.398	0.691
Behavioral × moderate crime	−0.032	0.466	−0.068	0.946
Behavioral × high crime	−0.388	0.963	−0.403	0.687
θ	0.1734			
Log-likelihood	−3 802.293			
Residual range	(−0.431,13.000)			

This table shows results similar to those found from the weighted least squares analysis from Chapter 4, but with easier interpretations. The only significant predictor is the number of insubordinate students. For every ten additional insubordinate students in schools that report suspensions, the number of suspensions reported is expected to increase by a factor of $e^{0.004 \times 10} \approx 1.041$. In other words, if a school reports 100 suspensions, a similar school with 10 more insubordinate students is expected to report $1.041 \times 100 \approx 104.1$, or around 104 suspensions.

6.5 Clickstream Analysis: Site Pairings

The English Wikipedia Clickstream data (Wulczyn and Taraborelli, 2015) are introduced in Chapter 1. We explore the link type popularity ranking by the number of pairings of referring and requested sites. Pairings of the referring and requested sites with ten or fewer observations were removed from the data set by Wikipedia analysts. The variables we use to model pairings are:

- n: the number of pairings of the referring and requested sites, i.e., the number of times a web path is used.
- Previous title: the result of mapping the referrer URL to the fixed set of values described above.
- Type: indicates if the pairings of the referring and requested sites are:
  - link: the referrer and request are both articles and the referrer links to the request.
  - redlink: the referrer is an article and links to the request, but the request is not in the production enwiki.page table.
  - other: the referrer and request are both articles but the referrer does not link to the request.

6.5.1 Exploratory Data Analysis

The EDA summary table (Table 6.13) of these clickstream data in Chapter 1 suggested that the counts of pairings of referrer and requested sites is a left-truncated, right-skewed distribution. The histograms of both the counts or the pairings and the natural log of the pairings indeed are left-truncated and right-skewed. The summary table and the normal Q-Q plot of the pairings counts and the log pairings counts is reproduced here for convenience in Table 6.13 and Figure 6.2.

Table 6.13 show that type and previous title are categorical variables. We saw in Chapter 1 that the pairings counts by category levels had no small counts or zeros. However, the combination of the pairings types and previous titles given in Table 6.13 shows zeros and counts less than 5 for several combinations of the levels, and hence, interacting these two categories as a predictor in a model is likely to fail as a model predictor. We therefore exclude this interaction.

6.5.2 Left-truncated Count Model

The clickstream data are left-truncated with only counts of 10 and larger. The data are the first approximately three hours of the data, resulting in 65 464 cases for analysis. We therefore must adjust the counts probability distribution to exclude the nonexistent data of less than 10 counts. To illustrate this with a simple probability count model, consider the Poisson probability distribution function we've discussed in Chapter 2 and earlier in this

Table 6.13 *English Wikipedia Clickstream continuous and categorical variables descriptive statistics. The number of pairings is the continuous (count) variable. Link type and previous title are the categorical variables.*

Variable	Minimum	Median	Mean	Maximum	Variance
Number of pairings	10	95.05	23	95 738	778 093.6

Variable	Levels	Number	Percent
Link type	Link	37 545	57.35%
	Other	27 602	42.16%
	Redlink	317	0.48%
Previous title	Bing	1 593	2.43%
	Empty	6 733	10.29%
	Google	8 078	12.34%
	Main_Page	509	0.78%
	Other	43 359	66.23%
	Wikipedia	3 962	6.05%
	Yahoo	1 231	1.88%

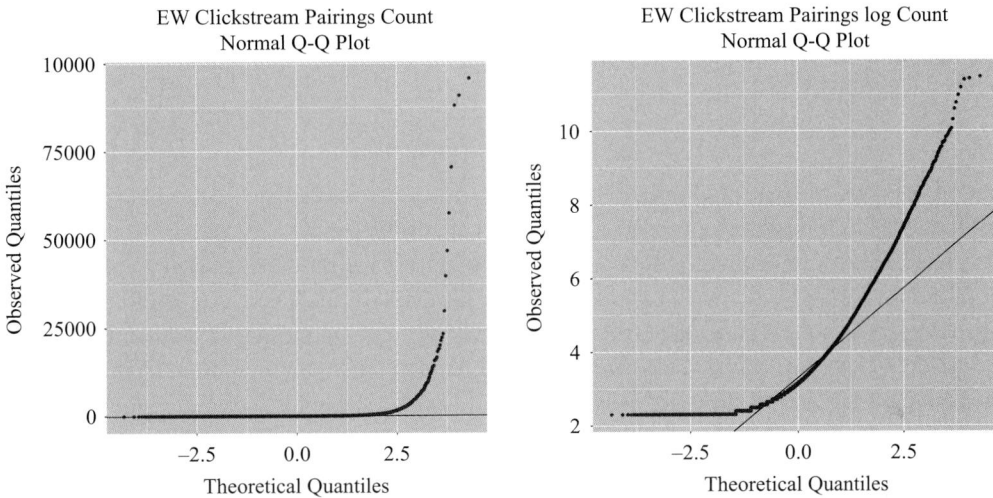

Figure 6.2 The normal Q-Q plots of the raw clickstream pairings counts (left-hand panel) and the natural log of the these counts (right-hand panel).

chapter. Recall

$$P(X=x) = \frac{\mu^x e^{-\mu}}{x!}, \tag{6.12}$$

where X is a random count which may take the values $x \in \{0,1,\ldots,\infty\}$ with probability determined from the Poisson probability function with mean and variance μ. The probability distribution function adjustment that can be used to describe when $x \in \{10,11,\ldots,\infty\}$ is to divide Eq. (6.12) by the sum of the probabilities that $X \in \{0,1,\ldots,9\}$, again with

Table 6.14 *English Wikipedia Clickstream left-truncated, weighted count model fit statistics comparison from worst (Poisson model) to best (Sichel 2).*

Model	Global deviance	df	*p*-value	AIC
Poisson	21 695 484	65 455	$\ll 0.001$	21 695 502
NB2	21 695 484	65 455	$\ll 0.001$	21 695 502
Delaporte	increasing, no model			
Sichel 1	599 754.9	65 453	$\ll 0.001$	5 997 76.9
Sichel 2	599 749.6	65 453	$\ll 0.001$	5 997 71.6

mean and variance μ, to correctly characterize the clickstream data (assuming these data follow a Poisson probability distribution). The same adjustment method is necessary for any probability distribution chosen to represent the clickstream data such as the negative binomial probability distribution function. Thus, for the Poisson distribution, we have that:

$$P(X = x \mid x > 9) = \frac{\frac{\mu^x e^{-\mu}}{x!}}{1 - \sum_{i=0}^{9} \frac{\mu^x e^{-\mu}}{x!}}. \tag{6.13}$$

Essentially, we must divide out the portion of the Poisson probability distribution function containing the probabilities of counts from 0 through 9. This truncated probability function scaling applies to whichever probability distribution we choose for the data set of interest.

6.5.3 Count Model Fit

The histogram of the clickstream data give no clear indication of what probability distribution functions may provide a best-fit probability model. As such, we constructed five possible models: truncated Poisson; truncated NB2; weighted truncated Delaporte; and weighted truncated Sichel with two different parameterizations. The weights are the number of same pairings in the data set applied to each pairing. The comparative evaluation of these five models (Table 6.14) indicates the weighted, truncated, second parameterization Sichel model is the best, and we therefore describe the model fit and analyze the effects. The Sichel distribution is particularly useful for highly overdispersed count data of the type exhibited by the clickstream data.

The X^2 test for the Sichel 2 model, as with all the models attempted, bespeaks a poor fit. This is confirmed in the left-hand panel of Figure 6.3 which shows the normal Q-Q plot of the left-truncated, weighted Sichel 2 model for the clickstream data. The obvious lack of fit appears in the upper quantiles of the Q-Q plot (right-hand panel) as the residuals deviate from the normal quantiles beginning at +3. Note that the model residuals are nearly homogeneous across the model fitted values as depicted in the right-hand panel of Figure 6.3.

Clickstream data are drawn from server logs, and hence each referrer and requested site pair is date and time stamped. The data set used in this analysis did not contain these stamps, and hence we are unable to test a time effect. However, under the assumption that the data are sequenced by the time of appearance, and with the understanding that the date and time stamps may have nonconstant intervals with possible missing data, a casual test for

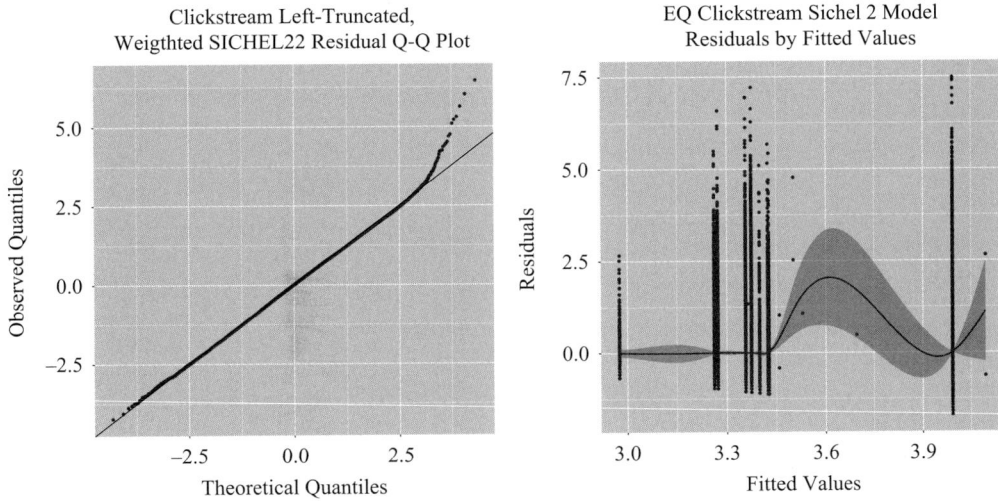

Figure 6.3 The left panel is the normal Q-Q plot of the clickstream pairings counts Sichel 2 model residuals. The right panel are the Sichel 2 model residuals as a versus the model's fitted values. The residual variance is nearly homogeneous across the fitted values.

autocorrelation of the sequence resulted with a nonstationary sequence. A first difference transformation produced a stationary sequence, and a first order moving average relationship was revealed. This suggests that these data be analyzed as a time series if the date and time stamps are available. The series is, of course, a sequence of counts evincing the use of a hidden Markov chain time series model. Unfortunately, this model type is outside the scope of this handbook.

While the model fit overall is inadequate, the Sichel 2 model gives only 0.58% of predicted counts under 10. There are only 0.09% predicted counts fewer than 8. This is an encouraging outcome considering the model construction data have no counts below 10, which the normal linear regression failed by fitting pairings counts beginning at 20.

6.5.4 Coefficient Interpretations

As a point of interest, we present the model effects in Table 6.15. We see that the Sichel 2 distribution parameters each are significant at any reasonable level. This property tells us that the location (μ), and shape parameters (σ and ν), give a reasonable fit to the clickstream count data.

The only three levels of any of either of the two categorical variables are the previous title levels of "empty," "Main_Page," and "Yahoo." A tactic often used in constructing effects models is to try various combinations of the previous title levels such as combining the three nonsignificant levels, or combining these three with any of the other levels. We don't pursue this technique here.

We tried several truncated count models to describe the English Wikipedia Clickstream pairings of referring and requested sites as a function of link type and previous title. None of the models gave desirable fit statistics, and further investigation is needed to find a

Table 6.15 *English Wikipedia Clickstream left-truncated weighted Sichel parameter estimates.*

Predictor	Estimate	Std. error	t-value	p-value
Link type level estimates				
μ intercept	3.828	0.272	14.050	0.000
Other	−0.561	0.130	−4.321	0.000
Redlink	−3.303	0.239	−13.825	0.000
Previous title level estimate				
Empty	−0.256	0.224	−1.147	0.251
Google	1.835	0.236	7.776	0.000
Main_Page	−0.134	0.379	−0.354	0.724
Other	−0.936	0.231	−4.047	< 0.001
Wikipedia	−1.089	0.243	−4.481	0.000
Yahoo	0.028	0.301	0.094	0.925
Additional Sichel parameters				
σ intercept	3.016	0.060	50.328	0.000
ν intercept	−0.848	0.013	−63.088	0.000

more appropriate model. However, our analysis did provide an introduction to the use of truncated count modeling while showing a clear improvement over the use of a normal linear model.

6.6 Summary

This chapter opened with a brief description of the various count model types including what conditions best suit the various count model types, measures of the model-to-data fit, and how to interpret the effects influencing the the response. We applied these several model types to the fire-climate, school survey, and clickstream data sets with counts of fires by decade, number of suspensions in schools, and pairings of referrer and requested sites, respectively. The fire-climate data were well-fitted by a zero-inflated, negative binomial model constructed from a Poisson-gamma mixture (NB2), after first trying variations of Poisson regression, negative binomial regression, and hurdle models. The effects of the NB2 predictors on the counts of fires were then analyzed.

A further application of the hurdle model was made on the numbers of annual suspensions in the SSOCS data. The hurdle model outcomes show results similar to those found from the weighted least squares analysis from Chapter 4, but with easier interpretations. Ease of interpretation is a very desirable model characteristic and hence the hurdle model is preferred over the weighted least squares analysis for this reason, and for using count models specifically designed for these count data.

The clickstream referrer and requested sites pairings count model, at best enjoyed a marginal fit by a Sichel 2 distribution generalized additive model. However, even though the fit was not adequate, we showed its superiority to the normal linear regression model, and pointed to alternative model trials such as the hidden Markov chain time series model for count time series for further investigation.

The takeaway from applying these count models to the fire, suspensions, and clickstream data is that the selected models were superior to the commonly used normal linear regressions demonstrated in Chapter 3.

6.7 Further Reading

We suggest the count modeling book of Hilbe (2014) for an excellent study the variety of count models with application examples of each. Specifically for negative binomial models, Hilbe (2011) is comprehensive and perhaps the most complete text on negative binomial modeling applications.

A staple of categorical data analysis, including modeling counts, is Agresti (1998). The seminal text on generalized linear models for count data is McCullagh and Nelder (1989).

A nice view of count regression models from the perspective of econometricians is given by Cameron and Trivedi (2013).

We used generalized additive models for the clickstream data and for more information on these models, see Hilbe (2014), Wood (2006). Another comprehensive reference for generalized linear models and general additive models is Faraway (2006).

7

Time-to-Event Response Models

7.1 Time-to-Event Data

Time-to-event (TTE) data are used to describe the probability of an event occurring by some specified time. TTE probability differs from logistic regression as the binary response analysis is focused on determining the probability of the event any time in the study. Commonly studied TTE data include the probability of patients in a cohort to be diagnosed with a hospital-related infection following a surgery, or the probability of automobiles failing a roadworthiness test. Directly related to TTE analysis is survival analysis which is the proportion of subjects or units surviving past a specified time. Two examples of survival analysis are the proportion of patients surviving a medical procedure, and the proportion of relay switches still functioning after an electrical current surge on a power grid. These four examples are events that are dependent on time. An example of an event that is not time dependent but still a function of an ordered sequence is how many meters of mass storage magnetic tape pass over the read/write head of the tape drive before a fatal fault is detected. In each of these examples, an event of interest is a function of an ordered sequence, either time or distance.

As TTE data are dependent on an ordered sequence, it is critical to identify disturbances in the sequencing of the events. A school system wishes to administer a curriculum designed to enable passing a sequence of calculus exams to qualified students, but several of the selected students have already passed at least the first of the exams prior to administration of the curriculum. A patient in a drug protocol study enters the study, experiences the desired response to the drug, then exits the study prior to the study end. A machine part fails in the time between two scheduled, routine inspections, resulting in an indeterminate exact time of the failure. A patient enters a study but withdraws prior to either the event of interest or the termination of the study. Finally, a student has no absences in a semester used for finding the time to an absence. Each of these examples of disturbance relative to a TTE sequence is a form of censored data.

The calculus test data contain left-censored data with the students who passed at least one exam prior to entry into the study, making the time to passing the exams indeterminate. The early exit of a patient from a study after responding to the drug is an example of right-censored data. The machine part failing between the two inspections is known as interval-censored data. The two examples of the subjects not experiencing the events of interest within the study span are considered censored, and they are called missing data in some texts, but are also known as right-censored data. Regardless, each is an example of a disturbance to the sequence and the TTE analysis must account for them. In addition, TTE analyses must account for staggered start and exit times, and studies designed to terminate after a specified number of events occurring or at a specified time.

TTE analysis differs from generalized linear model analysis, including normal linear regression, in that TTE is dependent on the ordered sequence whereas linear models are not. TTE analysis also must account for censored data while linear models often simply ignore them. Hence, TTE analysis is completely dependent upon the sequence.

7.2 Time-to-Event Models

Two TTE analysis strategies use either probability distribution functions such as the exponential, Weibull, or log-normal distributions, or the analysis is based on on an empirically constructed distribution function. The three probability distribution functions, among which there are many others, are often used in failure analysis for reliability studies. The empirical functions, often called nonparametric distributions, are commonly used in survival analysis, though these do not preclude the use of the named probability distribution functions. This chapter focuses on nonparametric distribution functions. Regardless of which distribution function is chosen or formulated, it is used to construct probability-based functions such as the cumulative distribution function, the survival function, and the hazard function.

The cumulative distribution function is formulated from a distribution function and measures the probability that some proportion of subjects or units in a study will experience the event of interest before a specified time t:

$$F(t) = P(T \leq t), \tag{7.1}$$

in which T is the time a proportion of the population of subjects or units experiences an event of interest, and t is the specified time chosen to see if the proportion has experience the event. $F(t)$ is the cumulative distribution function notation that represents the probability of occurrence.

The survivor function gives the probability of a portion of a population surviving past a specified time. The survivor function is obtained by subtracting each value in the cumulative distribution function from one, giving

$$S(t) = 1 - F(t), \tag{7.2}$$

where $S(t)$ is the survivor function. In reliability analyses, the survivor function is usually called the reliability function.

The hazard function represents the instantaneous rate of an event occurring within a given small interval of time after having no event to the beginning of this interval. The hazard function may be written as

$$h(t) = \frac{f(t)}{1 - F(t)}, \tag{7.3}$$

where $f(t)$ is the probability density function from which the cumulative probability function is constructed. The fraction represents the rate of nonsurvivors or failures in the specified small time interval.

A life table is a model that gives the probability of a subject not surviving or a unit failing within specified time intervals. They are derived directly from the cumulative probability distribution function. Life tables are constructed using fixed time intervals. If a study has a large number of subjects, shorter life tables may be constructed for specific combinations of, say, patient sex, age range, and some measure of overall state of health. The tables give straightforward access into the survival of particular population sample segments. Life tables are a convenient time-grouping technique for survival data as the interval definition for the table is particular to the study needs.

The Kaplan-Meier (K-M) method does not group subject entry times into intervals as do life tables, but rather uses each subject's time of the event of interest to produce survival statistics. For data sets with no censoring, both the K-M analysis and the life table analysis result in the same survival statistics. When censored data are present, the K-M method is used to account for the various forms of censoring, thus giving survival statistics that are unbiased.

The Cox proportional hazards model is a semiparametric approach to describing the relationship of predictors to survival times. It is semiparametric in that the Cox method does not require the use of a designated probability distribution to model survival times, and thus may be empirically constructed. The basic Cox regression equation is

$$h_i(t) = h_0(t) \exp(\beta_0 + \beta_1 x_{i1} + \cdots + \beta_p x_{ip}), \tag{7.4}$$

where $h_0(t)$ is known as the baseline hazard for a subject or unit for which the predictors all have values of zero. For subject or unit i, the hazard $h_i(t)$ is an anti-log linear function of p predictors. In Eq. (7.4), the x_{ij} represent observed predictor values, while the β_j are unknown coefficients that must be estimated. The baseline hazard $h_0(t)$ is typically not estimated, as interpretations are made with respect to ratios of hazards.

The Cox proportional hazards model in Eq. (7.4) is a proportional hazards model as the ratio of any two subjects i and j is fixed such that hazard $h_i(t)/h_j(t)$ is a constant for all times t. The plotted hazards of individuals or units form curves that everywhere in time have constant distances between all possible, same time combinations of curve points. Nonproportional hazards versions of the Cox model can be used when appropriate.

Sample size in proportional hazards analysis dictates how may predictors may be tested. A sample size of approximately 60 allows for 5 or fewer predictors while sample sizes greater than 60 allow for more than 5 predictors. When the predictors are multivariate normal, linear with respect to the log of the hazards proportion, the power of the Cox proportional hazard analysis is optimal. EDA of the predictors will aid in identifying conformance or deviations from these conditions. TTE analyses are sensitive to correlations among the predictors. High standard errors for estimated parameters suggest predictor correlation. EDA pairwise correlation tables give clues to predictor pairwise correlation, and a multiple linear regression may be useful for identifying multicollinearity among the predictors.

To use a Cox proportional hazards model, the shape of the hazard proportions must be the same throughout the study period for all observations and all study groups as a consequence of the fixed curve distances for each time value. This fixed distance may be tested as a predictor dependence on time analysis. Once events begin, the rate of occurrence must proceed at the same rate throughout the study for all the study groups. A general Cox regression model relaxes the requirement for constant proportional hazard, thus permitting the use of time-dependent predictors; i.e., often, predictor values or levels change with time, and the Cox regression model may be used.

The Cox model fit assessment includes the analysis of the model residuals, specifically the use of Schoenfeld residuals. The Schoenfeld residuals (either overall or for an individual predictor) can be thought of as the difference between predictor values at each of the failure times and a weighted average of predictor values at all times. General Schoenfeld residuals average these over all the predictors and test for individual predictors by making calculations for a single predictor at each time value. Plots of these residuals against time are examined to see if they show evidence of a trend. If a trend suggests this difference changes over time, a time-dependent predictor may be responsible.

We use the Framingham Heart Study data to demonstrate the use of the life tables, Kaplan-Meier method, and the Cox proportional hazards model. In addition, we compare the outcomes of these methods to demonstrate differences.

7.3 FHS Analysis: Time to Hypertension

We wish to study which variables from the Framingham Heart Study influence the time to and from hypertension (TIMEHYP), given no known previous hypertension. For example, does the number of cigarettes used per day shorten the time to an onset of hypertension? The variables we use to model survival of hypertension are:

- SEX: sex of the participant
- AGE: participant age in years
- CIGPDAY: number of cigarettes smoked per day
- TOTCHOL: serum total cholesterol (mg/dL)
- PREVHYP: prevalent hypertensive
- HYPERTEN: diagnosed in the study period as hypertensive
- TIMEHYP: time to first diagnosis of hypertensive

The FHS data first had all cases in which a subject was diagnosed as hypertensive prior to the study start date removed. Any cases for which values of the variables listed above were missing were also removed. Table 7.1 is a summary of the continuous and counts variables included in the TTE analysis.

Table 7.2 gives the numbers and proportions of the sex levels (1 = male, 2 = female), subjects' level of diabetes (0 = no diabetes, 1 = diabetes present), and subjects' hypertensive state (0 = no hypertension, 1 = hypertensive).

Both the continuous and dichotomous variables are suspected of contributing to the time to hypertensive diagnosis within the study span. We will begin with an examination of survival with life tables (Section 7.3.1) and plots of life tables which are known also as survival function plots. These plots include examinations for group differences. The final analyses are using the Cox proportional hazards model and Cox regression.

7.3.1 Life Tables

Table 7.3 is the life table for one year intervals (year) with the number of subjects entering each year interval who are not hypertensive. The number of subjects per interval who

Table 7.1 *Framingham Heart Study descriptive statistics for continuous variables.*

Variable	Minimum	Median	Mean	Maximum	Variance
Time of hypertension	0	2429	3599	8766	11996207.990
Total cholesterol	107.0	238.0	241.2	696.0	2059.309
Age	32.00	54.00	54.79	81.00	91.325
Cigarettes	0.00	0.00	8.25	90.00	148.000

Table 7.2 *Framingham Heart Study descriptive statistics for categorical variables.*

Variable	Levels	Number	Percent
Hypertension	Yes	8642	74.3%
	No	2985	25.7%
Sex	Female	6605	56.8%
	Male	5022	43.2%
Diabetes	Yes	530	4.6%
	No	11097	95.4%

are lost to the study for any reason such as death is in the "number attributed" column. The table has the number of subjects in each interval who are at risk (number at risk) of becoming hypertensive. The number of subjects who became hypertensive is in the "number of diagnoses" column. The life table estimated survival function value is given for each interval in the "proportion surviving" column. The study spans 24 years, so the final interval represents the status of subjects surviving (not becoming hypertensive) after 24 years. The life table suggests approximately 45% of subjects do not become hypertensive. The reasons, however, are not revealed by this table.

The table is interpreted for, say, the 8–9 years interval as 1 949 subjects enter this year's interval without hypertension. 27 subjects are lost to the study in this period, giving 1 936 subjects (an estimated mid-year value) at risk of becoming hypertensive. Of these at risk, 67 subjects are identified as hypertensive. The estimated proportion surviving without being hypertensive as of this interval is approximately 0.79, i.e, 79% of the study subjects remaining by year 8–9 have not been diagnosed with hypertension.

The left-hand panel of Figure 7.1 is a plot of the life table estimated survival function, the proportion surviving, versus the time in yearly intervals. Note the stair-step characteristic of summarizing by time to event intervals with preset, year-sized spans. The plot may be read in two ways. The first is by specifying the proportion surviving on the vertical axis, projecting this proportion to the survival curve, and then drawing down from the survival curve to read the years to which this proportion survives. The second method is to reverse the process of the first method by establishing a time on the horizontal axis, reading up to the survival curve, and then reading from the survival curve to the vertical axis for the proportion surviving.

Table 7.3 *Framingham Heart Study life table for diagnosis of hypertension.*

Year	Number of subjects	Number attritions	Number at risk	Number of diagnoses	Proportion surviving
0–1	2 932	4	2 930.0	6	1.0000000
1–2	2 922	54	2 895.0	89	0.9979522
2–3	2 779	70	2 744.0	107	0.9672725
3–4	2 602	67	2 568.5	79	0.9295545
4–5	2 456	62	2 425.0	87	0.9009640
5–6	2 307	54	2 280.0	92	0.8686407
6–7	2 161	43	2 139.5	59	0.8335903
7–8	2 059	48	2 035.0	62	0.8106028
8–9	1 949	27	1 935.5	67	0.7859063
9–10	1 855	48	1 831.0	81	0.7587011
10–11	1 726	26	1 713.0	31	0.7251376
11–12	1 669	39	1 649.5	59	0.7120148
12–13	1 571	25	1 558.5	40	0.6865472
13–14	1 506	34	1 489.0	36	0.6689264
14–15	1 436	29	1 421.5	46	0.6527536
15–16	1 361	30	1 346.0	47	0.6316304
16–17	1 284	34	1 267.0	50	0.6095749
17–18	1 200	19	1 190.5	26	0.5855191
18–19	1 155	25	1 142.5	47	0.5727316
19–20	1 083	14	1 076.0	32	0.5491707
20–21	1 037	27	1 023.5	47	0.5328385
21–22	963	21	952.5	33	0.5083701
22–23	909	19	899.5	34	0.4907572
23–24	856	24	844.0	36	0.4722072
24–NA	796	310	641.0	486	0.4520657

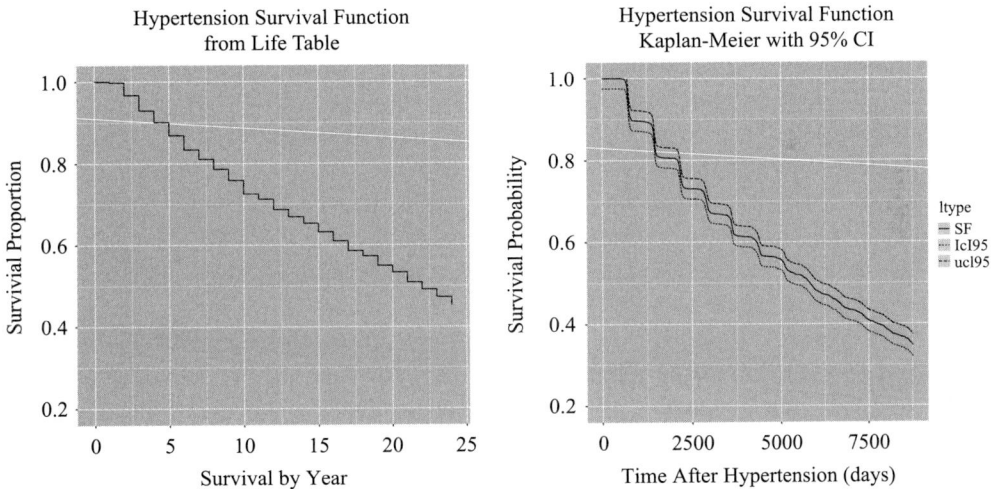

Figure 7.1 Life table estimated survival function (left panel) by yearly intervals with no censoring. The right-hand panel is the Kaplan-Meier survival function by day rather than by year, and includes censoring. The proportion of survivors scale (vertical axis) for the two plots are the same.

Table 7.4 *Framingham Heart Study Kaplan-Meier survival table for diagnosis of hypertension.*

Days	Day of diagnosis	Number at risk	Number of diagnoses	Number of censored	Proportion surviving
1	0	2 932	1	0	1.000
2	538	2 919	1	0	0.999
3	554	2 918	1	0	0.999
⋮	⋮	⋮	⋮	⋮	⋮
592	3 710	1 693	1	0	0.617
593	3 719	1 692	1	0	0.617
594	3 725	1 691	2	0	0.616
⋮	⋮	⋮	⋮	⋮	⋮
1 185	8 756	797	1	0	0.348
1 186	8 761	796	1	0	0.348
1 187	8 764	795	1	0	0.347

7.3.2 Kaplan-Meier Method

The Kaplan-Meier (K-M) method does not group patient entry times into intervals as in life tables, but rather uses each patient's time of event (diagnosis of hypertension) to produce survival statistics. Note that the date is the date of a diagnosis even though the hypertension onset likely occurred between two examinations, making the onset of hypertension interval-censored data. With no censoring on the time to diagnosis, both the K-M analysis and the life table analysis in the previous section result in the same survival statistics.

Table 7.4 is a shortened display of the full K-M life table of the FHS time-to-hypertensive data. The days until hypertension diagnosis is given in the "days of entry" column of the table. The number of subjects at risk for the corresponding time is given in the "number at risk" column and the number of subjects lost to the study at any given time is in the "number of censored" column. The "proportion surviving" column gives the estimated survival function proportion.

We read Table 7.4 as follows: for the 592nd patient that received a diagnosis of hypertension; the diagnosis was made 3 710 days into the study. The number of at-risk patients at that time was 1 693. Only one patient was diagnosed with hypertension that many days into the study and no patients were lost to the study at that moment. The proportion of patients at-risk of hypertension 3 710 days into the study was approximately 0.617. Therefore 61.7% of the patients still in the study are at risk of being diagnosed as hypertensive. Note that this K-M analysis does not account for why a patient may be diagnosed as hypertensive.

Figure 7.1 gives a plot of the K-M survival function in the right-hand panel. Notice that the life table survival curve is less concave than the K-M curve as the former is based on annualized intervals whereas the latter is the actual number of days until hypertensive diagnosis. Also, unlike the life table method, the K-M method accounts for censored subjects; i.e., subjects that were lost to the study. The proportion scale is the same for the life table and the K-M table. The plots clearly show the difference between the proportions surviving without censoring (annualized time interval life table) and with censoring (K-M table).

Sex and the presence of diabetes are dichotomous predictors. The K-M method (as do life tables) allow us to compare the time to hypertension diagnosis of the two levels of each of these variables. The first row panel of Figure 7.2 displays the difference in survival functions between males and females. We read the plot as between 1 000 and 2 000 days, males and females diverge with the proportion of females being not hypertensive less than that of males. The trend maintains until approximately 7 000 days when the females not becoming hypertensive proportions approximately match that of males.

The second row left panel of Figure 7.2 shows the survival function difference between having and not having diabetes. The survival functions for diabetes clearly separate at

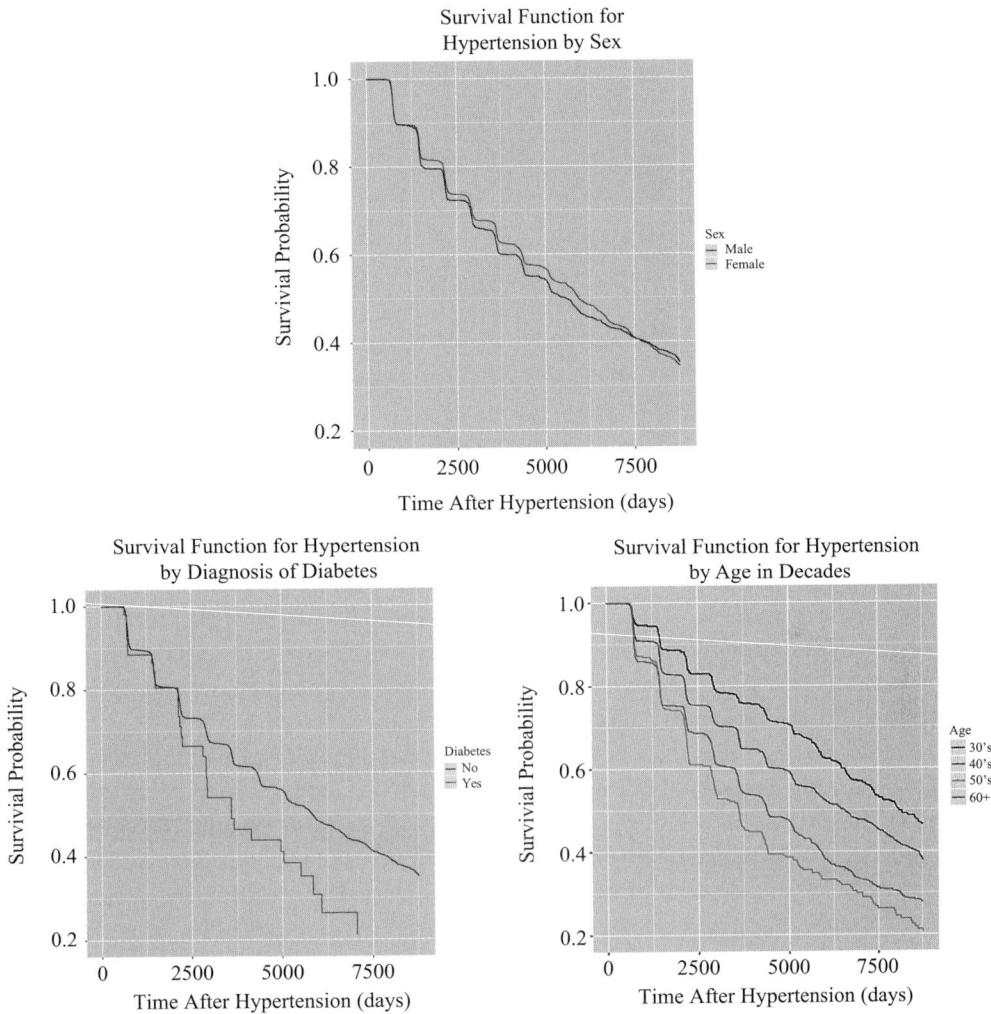

Figure 7.2 K-M survival curves of hypertensive diagnosis by sex (left-hand panel), for patients with diabetes (center panel), and by age in decades. The survival probability scales for each are the same.

Table 7.5   *Framingham Heart Study predictor summary. The p-values for both the skew and kurtosis statistics indicate that the probability distributions of these predictors are skewed and have tails that deviate from those of a Gaussian probability distribution.*

Predictor	Skew	Skew z-value	Skew p-value	Kurtosis	Kurtosis z-value	Kurtosis p-value
Cholesterol	0.516	11.438	$\ll 0.001$	0.531	5.878	$\ll 0.001$
Age	0.429	9.5116	$\ll 0.001$	-0.790	-8.752	$\ll 0.001$
Cigarettes	1.124	24.909	$\ll 0.001$	0.708	7.847	$\ll 0.001$

just over 2 000 days when the proportion of diabetic patients being not diagnosed with hypertension dramatically decreases over patients without diabetes.

We segmented age from years into decades, and the corresponding survival curves plotted in the second row right panel of Figure 7.2. The decades of patients in the study are those in their 30s, 40s, 50s, and then 60s and older. Not surprisingly, the proportion of patients having diagnosis of hypertension of successively older patients decreases faster by increasing age category.

7.3.3 Cox Proportional Hazards Models

The Kaplan-Meier table and life table, and corresponding plots, suggest dependency of time to diagnosis of hypertension on sex, having diabetes, and age, though there may be other time-dependent predictors. We wish to test the effects of sex, age, number of cigarettes smoked per day, presence of diabetes, the total cholesterol, the interaction of sex with the number of cigarettes per day smoked, and the interaction of sex and the presence of diabetes on the occurrence of hypertension. However, we are not concerned with the probability of being hypertensive as with logistic modeling, but rather on the length of time until a patient is diagnosed with hypertension. In addition to estimating the effects of these predictors, we wish to account for censored data. The Cox proportional hazards (CPH) regression model accommodates these conditions. However, we must check the CPH model assumptions prior to using it. The assumptions we must address are sample size, normality (Gaussian distribution), outliers, and multicollinearity.

The sample size is more than adequate for the seven predictors (sample size must be greater than 60) after removing cases having missing data or hypertension at study entry resulting in 2 946 patients. Examination of Table 7.5 suggests the three continuous or count predictors are skewed with thick tails. Figure 7.3 shows Q-Q plots of these three predictors before (top row) and after (bottom row) a natural-logarithm transformation and square-root transformation for the number of cigarettes used per day. Only cholesterol shows improvement, so we use the ln(cholesterol) transformation in the proportional hazards model. Other transformations such as Box-Cox transformations may be tried, but interpretation should be a concern in the selection.

Table 7.5 and Figure 7.3 show that the log of total cholesterol, age, and cigarettes used per day have significant skew (large z-values and p-values $\ll 0.001$) which is an

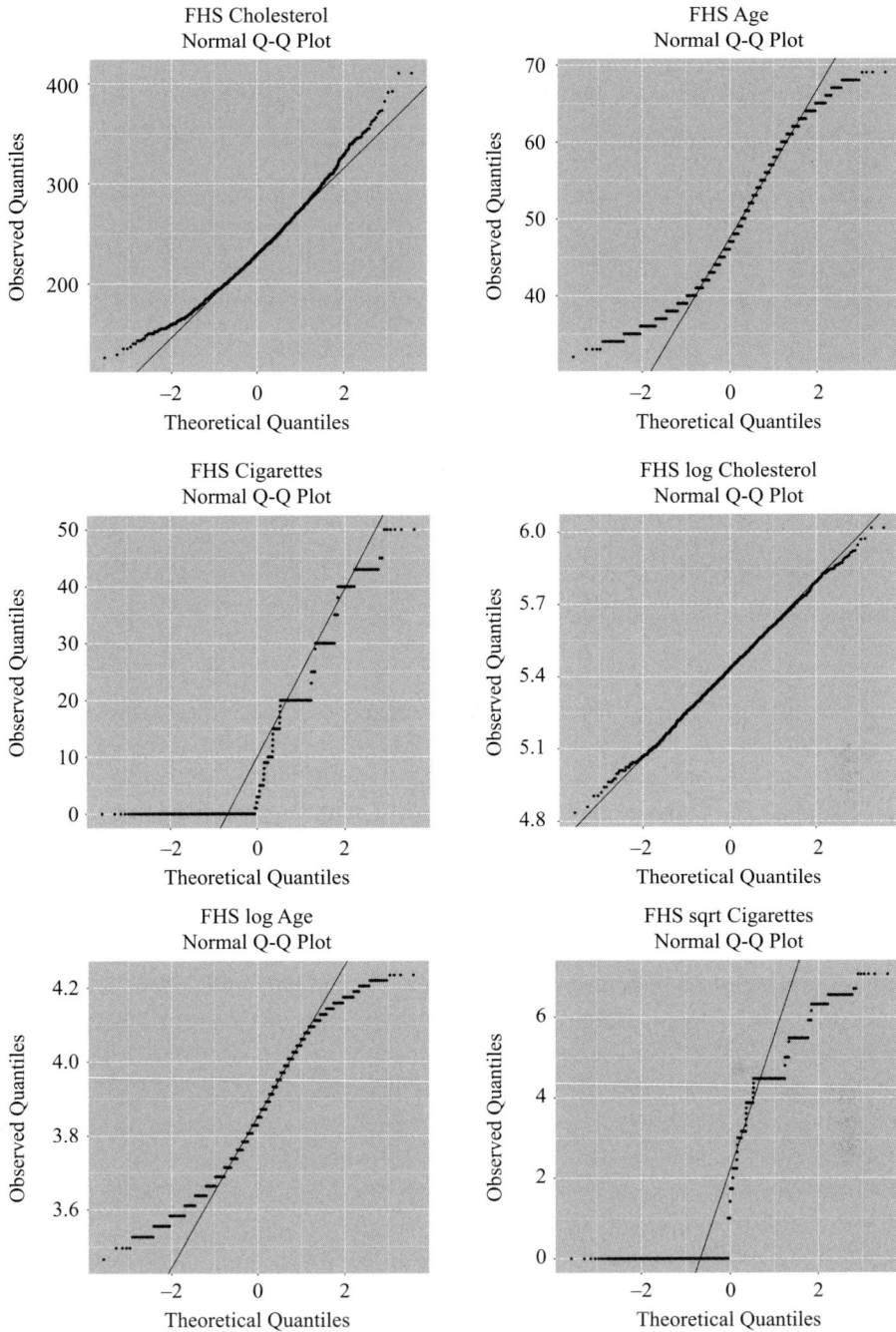

Figure 7.3 Q-Q plots of total cholesterol, age, and the number cigarettes used per day in the first row left-hand panel, center panel, and right-hand panel, respectively. The second row are the Q-Q plots of the log of the total cholesterol (left-hand panel), log of age (center panel), and the square root of the number of cigarettes used per day (right-hand panel).

Table 7.6 *Framingham Heart Study predictors with
excessive Mahalanobis distance cases summary.*

Variable	Minimum	Median	Mean	Maximum
log cholesterol	1.553	1.680	1.672	1.815
Age	37	48	50.00	67
Cigarettes	0.00	60	43.21	60

Table 7.7 *Framingham Heart
Study predictor variance inflation
factors (VIF). The VIF values
much less than 10 suggest no
multicollinearity.*

Variables	VIF
log cholesterol	1.1016
Age	1.1244
Cigarettes	1.0235

indication of possible outliers. The Mahalanobis distance is calculated to ascertain if these predictors exhibit as multivariate outliers. We use the Mahalanobis distance as it measures how far from a multivariate normal distribution a combination of predictor values lies from the distribution. Using a detection criterion of 0.001 and 3 degrees of freedom (χ^2 = 16.266), 14 (0.48% of the 2 946 patients) cases were identified as potential multivariate outliers. The summary values of these multivariate outlier cases is given in Table 7.6. Each case is further tested by creating an indicator variable response for each case. Thus 14 regressions for the 14 suspected multivariate outlier cases are examined to see which and how often the three predictors have significant coefficient estimates. The regressions give 4 occurrences of log total cholesterol, 4 occurrences of age, and 12 occurrences of number of cigarettes used per day which, after accounting for cases occurring in 2 or 3 of the predictors, all 14 cases were identified as multivariate outlier cases. These 14 cases are removed, leaving 2 932 patients in the analysis. The cases are removed so as to not decrease the power of the Cox proportionality hazards analysis for purposes of demonstration. However, these cases must be examined carefully before removing them relative to the study objectives.

Another check of Cox regression model assumptions is for multicollinearity among the predictors. Table 7.7 shows the variance inflation factor (VIF) values for the log of the total cholesterol, age, and the number of cigarettes used per day. Each VIF value is much less than 10 thereby suggesting no multicollinearity is present among these predictors. This finding is what may be deduced from the FHS scatter plot in Chapter 1.

The logistic model of Chapter 5 tested the interaction between sex and diabetes status, and the interaction between sex and the number of cigarettes used per day on the probability of hypertension diagnosis. The outcome of the logistic model showed that the interactions were not significant contributors to the probability of diagnosis of hypertension. However,

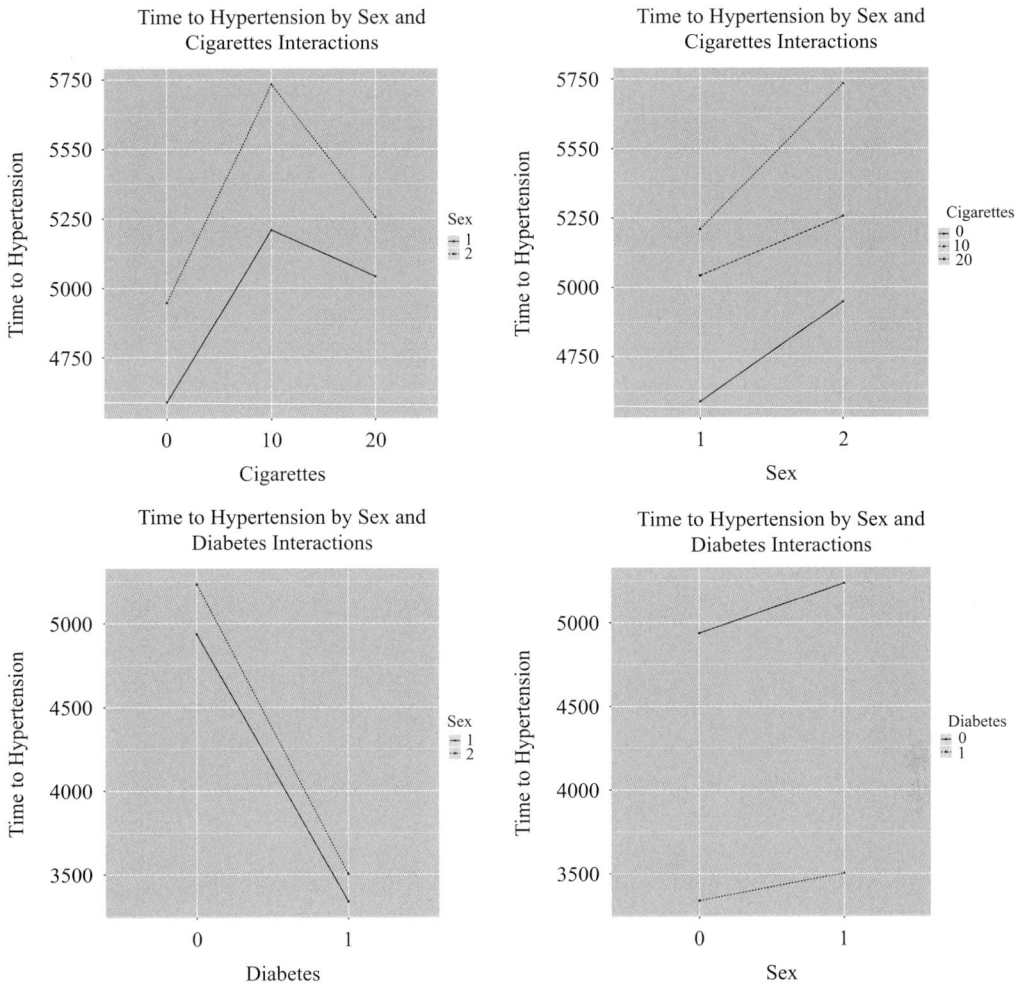

Figure 7.4 Interactions between sex and the number of cigarettes smoked per day and the presence of diabetes. The lack of parallel lines in each panel suggests these predictors interact.

Figure 7.4 depicts the interaction plots using the time to diagnosis of hypertension by sex and cigarettes per day, and sex and diabetes status. The panels of the plot show nonparallel lines for each interaction which suggest sex influences the time to hypertensive diagnosis depending on the level of diabetes status. Similarly, sex influences the time to hypertensive diagnosis depending on the level of cigarettes used per day. Therefore we include this interaction in the CPH model to test the significance and strength of the interaction.

Table 7.8 gives the fit statistics of the CPH model using only main effects. The main effects CPH model makes no assumption as to whether each predictor is dependent on the time. If the predictor levels are dependent on time, then this will change the time to hypertensive diagnosis. We include the time-independent predictor CPH model to compare with the time-dependent predictors Cox regression model to show how the time dependence

Table 7.8 *Framingham Heart Study Cox Proportionality Hazard model parameter estimates of time to diagnosis.*

Predictor	Estimate	exp(estimate)	Std. error	z-value	p-value
Age	0.027	1.028	0.003	8.803	$\ll 0.001$
Cigarettes	0.008	1.008	0.007	1.110	0.267
log cholesterol	0.464	1.590	0.140	3.303	0.001
Diabetes	−0.047	0.954	0.538	−0.087	0.931
Sex	0.076	1.079	0.065	1.167	0.243
Cigarettes × sex	−0.008	0.992	0.005	−1.671	0.095
Diabetes × sex	0.215	1.240	0.356	0.604	0.546

Table 7.9 *Framingham Heart Study test for nonproportional hazard of time to diagnosis.*

Predictor	ρ	χ^2	p-value
Sex	0.049	4.298	0.038
Age	−0.0485	4.3287	0.0375
Cigarettes	0.039	2.995	0.084
Diabetes	0.047	4.047	0.044
log cholesterol	−0.081	11.680	0.001
Global		34.0306	0.0000

changes the adequacy of these models. The CPH AIC = 26 542.99 which, along with the model fit statistics, we shall compare to the time-dependent predictor Cox regression below.

The K-M plots of sex, diabetes and age show that we must examine whether each predictor interacts with the time to hypertension diagnosis used to construct the survivor function. Significant predictor interaction with time is an indicator that the predictor violates the assumption that survival rates have fixed proportion, and hence, not the same. In Table 7.9, the predictors are tested as cross products with the specified transformation of time to hypertensive diagnosis (natural log in this case). The scaled Schoenfeld residuals are in the same order as the predictors were entered in the Cox model. The correlation in the table is simply a linear correlation corresponding to a two-variable scatter plot of the Schoenfeld residuals versus the time to diagnosis of hypertension.

Table 7.9 gives the predictor in the first column, the correlation value of the predictor by the log of the time to hypertension diagnosis, the value of the χ^2 variate, and the associated significance of the χ^2 test as the p-value. The first element of the list corresponds to the scaled Schoenfeld residuals for sex (sex = 2 [female] rather than the reference level) which has a correlation value $\rho = 0.577$, $\chi^2 = 5.9785$ giving a p-value of less than 0.05. As we saw above in the K-M plot for sex, the hazard rates for the sexes are different at the 0.05 error rate level. The predictors sex, age, diabetes, and log total cholesterol test as nonproportional hazard predictors, and thus must be modeled as cross products with time.

Hypertension Predictor Parameter Estimates vs. Time

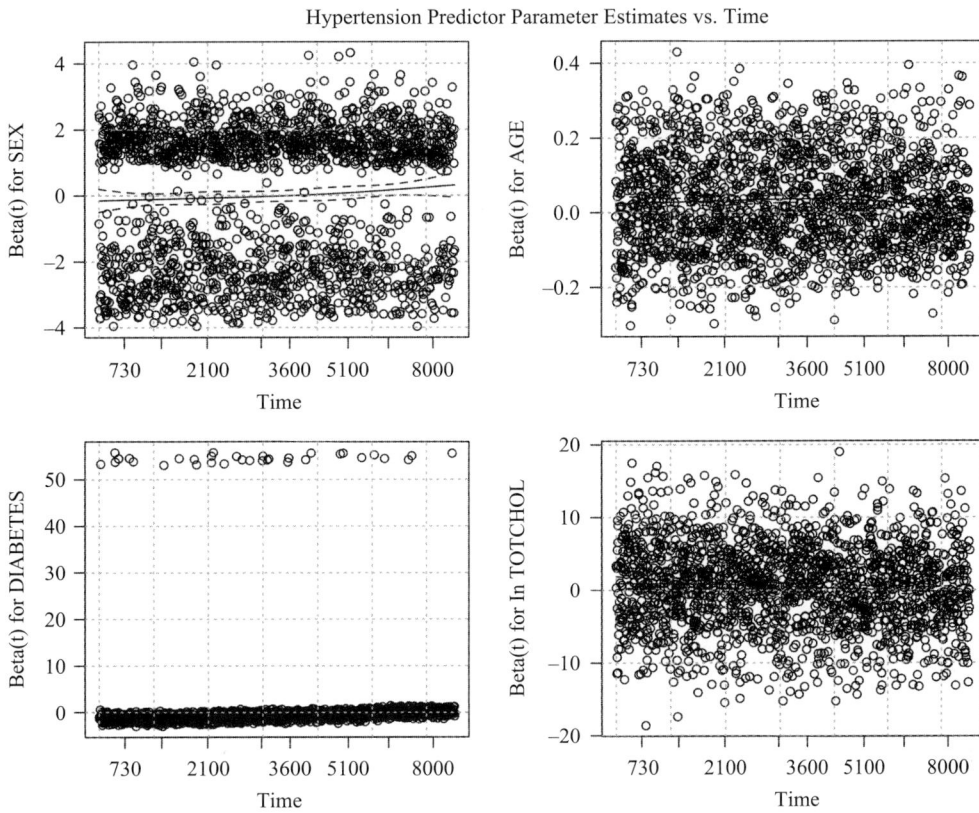

Figure 7.5 Nonproportional hazards test plots of sex, age, diabetes, and log total cholesterol. The plots suggest time dependence is not necessary.

Plots of Schoenfeld residuals may be made of each predictor's coefficient. The vertical axis is the coefficient estimate as a function of time, and the horizontal axis is the time to diagnosis of hypertension.

In each of the panels of Figure 7.5, the zero value of the coefficient is given as the reference. If the solid line representing the time value of the coefficients lies on or near the zero line, the predictor may be interpreted as not being dependent on time. The curves in each panel are smoothed (natural spline) fits of the time-varying estimates of the log of the hazard ratio. The curves for each predictor are relatively flat, straight, and have 95% confidence intervals containing the zero reference, contradicting the test for proportional hazard, i.e., cross products with time are not indicated. As the χ^2-test of total cholesterol is contradicted by the zero-proportions hazard plot, we suggest including the cross product in the model and evaluate the significance of this interaction in the presence of the other predictors.

7.3.4 Time-Dependent Cox Models

The tests of the time-dependent predictor assumptions and plots of the life table and K-M methods suggest a Cox regression that includes interactions between predictors, and between

Table 7.10 *Framingham Heart Study Cox proportionality hazards interaction model of time to diagnosis compared to the main effects. The 2LL column is the log-likelihood ratio statistic on 7 degrees of freedom for the CPH model, and time-dependent Cox regression model has 11 degrees of freedom.*

Model	2LL	Wald	Score	R^2	AIC
Time interaction	154.0	156.7	159.1	0.004	26 532.66
Main effects	135.6	139.0	140.6	0.003	26 542.99

predictors and time. The predictors are age, the number of cigarettes per day, the log total cholesterol, diabetes status, sex, diabetes crossed with sex, cigarettes per day crossed with sex, and the time interactions of sex, age, diabetes, and log total cholesterol. We first examined the goodness-of-fit statistics for this model and, by way of comparison, with a CPH model of age, cigarettes per day, log total cholesterol, diabetes, sex, sex by diabetes, and sex by cigarettes per day. Following the goodness-of-fit analysis, we evaluate the predictor influence on the time to hypertension diagnosis.

Time-dependent Cox models' goodness-of-fit is evaluated by comparing a model of interest to an intercept-only model, i.e., every Cox model may be compared to an intercept-only model having no other predictors. Three tests of the Cox model with predictors and without predictors are the likelihood ratio test, the Wald test, and the score test. The likelihood ratio test is a χ^2 test of the predictor model log-likelihood by the null model log-likelihood. These goodness-of-fit statistics test that the predictor coefficients are zero versus at least one coefficient is not zero. (Note this is not a deviance statistic as in logistic or count models as there is no saturated model.) The estimate ratio is 154.0 on 11 degrees of freedom giving a p-value much less than 0.05, which indicates the predictors estimate the time to hypertension diagnosis better than an intercept-only model. The Wald test estimate and the log rank score test estimate for each of the values with p-value much less than 0.05. Hence, all three tests indicate the predictor model is better than the intercept model. This is true of the CPH as well. As a point of interest, we compare the time interaction model with the goodness-of-fit for a main-effects-only model in Table 7.10. All the goodness-of-fit statistics for the time interaction model are greater than the goodness-of-fit statistics for the main effects except for the AIC. The first three statistics are described above and the time interaction model values of these indicate the time interaction model has a better fit. The CPH R^2 value is the improvement in likelihood between the fitted model and the intercept-only model, and thus should not be used to compare models. The time interaction $R^2 = 0.004$ suggests the improvement of the time interaction over the intercept-only model is marginal (maximum $R^2 = 1$). However, we advise not using the R^2 statistic as its interpretation is often confounded with the interpretation of the R^2 statistic of normal linear regression. Finally, using the AIC values of the two models, we see that the time interaction model AIC = 26 532.66 is less than the main effects AIC = 26 542.99 which indicates the time interaction model fits the FHS time to hypertension diagnosis marginally better than does the main-effects model. This outcome supports results drawn from the nonproportional hazards test plots.

Table 7.11 *Framingham Heart Study Cox regression model of time to diagnosis.*

Predictor	Estimate	exp(estimate)	Estimate std. error	z-value	p-value
Sex	−0.130	0.878	0.138	−0.945	0.345
log time × sex	0.0990	1.104	0.061	1.614	0.107
Age	0.036	1.037	0.008	4.462	≪ 0.001
log time × age	−0.005	0.995	0.004	−1.178	0.239
Cigarettes	0.007	1.007	0.007	1.048	0.294
Diabetes	−0.913	0.402	0.692	−1.319	0.187
log time × diabetes	0.471	1.601	0.246	1.915	0.056
log cholesterol	1.506	4.507	0.384	3.924	< 0.001
log time × log cholesterol	−0.514	0.598	0.178	−2.887	0.004
Sex × cigarettes	−0.008	0.993	0.005	−1.629	0.103
Sex × diabetes	0.218	1.244	0.357	0.611	0.541

Table 7.11 displays the estimates of the relative effect of each time-dependent Cox model predictor's coefficients including the cross products and the Wald tests of significance, viz., the z-value and the associated p-value. The estimated coefficient (estimate), the hazard ratio (anti-log of the coefficient estimate, exp(estimate)), and the coefficient estimate standard error (Estimate std. error) are given. The significant influencers (0.05 error rate) on the time to hypertension diagnosis are age and the log total cholesterol by log time.

The age predictor may be evaluated independently of time while the log total cholesterol effect on time to hypertension diagnosis may only be assessed in terms of the time interaction. As the log total cholesterol effect on time to hypertension changes over time and is influenced by how long a patient has been followed, the main effects CPH model presented above could give misleading results. The Cox regression model with time-dependent effects accounts for the conditions that confound the time to hypertensive diagnosis. Thus, the model is no longer a proportional hazards model. The time-dependent Cox regression model compares the risk of a hypertensive diagnosis between patients with hypertension and not-yet-diagnosed patient at each event time, and re-evaluates the risk group each patient belongs in based on diagnosis status by that time.

The time-dependent predictors are divided into annual at-risk intervals representing years to a hypertensive event. The days to diagnosis of hypertension is agglomerated into years with the first event occurring between the first and second years. At least one event occurs in each yearly interval through all the 24 years of the study. Note that not all of the patients in the study experienced an event meaning many patients at the end of the study had no event. The Cox regression model with the time-dependent predictors is given in Table 7.11. Table 7.12 is a life table for for age = 45 years, cigarettes per day = 10, log total cholesterol = ln(300), diabetes status = 1 (patient has diabetes), and sex = 1 (male). For example, consider the row in Table 7.12 for time = 9.0000. At 9 years, 1 865 patients remain at risk for a diagnosis of hypertension. In year 9, 12 patients were diagnosed with hypertension, leaving the proportion of patients surviving hypertensive diagnosis at 0.6352.

Table 7.12 *Framingham Heart Study Cox regression life table of diagnosis to hypertension.*

Time	n.risk	n.event	Survival	Std. error	Lower0.95CI	Upper0.95CI
1.0000	2927.0000	1.0000	0.9998	0.0003	0.9993	1.0000
2.0000	2918.0000	298.0000	0.9119	0.0319	0.8515	0.9766
3.0000	2607.0000	10.0000	0.9085	0.0151	0.8793	0.9386
4.0000	2590.0000	249.0000	0.8148	0.0285	0.7607	0.8727
5.0000	2322.0000	7.0000	0.8119	0.0229	0.7683	0.8580
6.0000	2292.0000	210.0000	0.7213	0.0324	0.6605	0.7876
7.0000	2067.0000	6.0000	0.7185	0.0301	0.6619	0.7801
8.0000	2045.0000	165.0000	0.6410	0.0375	0.5716	0.7189
9.0000	1865.0000	12.0000	0.6352	0.0364	0.5677	0.7107
10.0000	1841.0000	143.0000	0.5645	0.0433	0.4857	0.6561
11.0000	1676.0000	15.0000	0.5568	0.0424	0.4797	0.6464
12.0000	1650.0000	118.0000	0.4963	0.0482	0.4103	0.6003
13.0000	1516.0000	15.0000	0.4885	0.0470	0.4045	0.5900
14.0000	1486.0000	90.0000	0.4412	0.0511	0.3516	0.5537
15.0000	1382.0000	33.0000	0.4237	0.0511	0.3345	0.5368
16.0000	1334.0000	84.0000	0.3793	0.0548	0.2858	0.5034
17.0000	1230.0000	34.0000	0.3612	0.0547	0.2684	0.4860
18.0000	1180.0000	60.0000	0.3292	0.0564	0.2352	0.4606
19.0000	1107.0000	28.0000	0.3142	0.0558	0.2219	0.4450
20.0000	1063.0000	55.0000	0.2849	0.0570	0.1925	0.4216
21.0000	995.0000	32.0000	0.2679	0.0565	0.1772	0.4050
22.0000	943.0000	50.0000	0.2415	0.0570	0.1520	0.3836
23.0000	879.0000	27.0000	0.2274	0.0560	0.1403	0.3685
24.0000	836.0000	36.0000	0.2086	0.0555	0.1239	0.3513

The proportion surviving is contained by the 95% confidence interval of 0.5677 on the low end to 0.7107 on the high side.

Figure 7.6 displays the survival curves for the Cox proportional hazards model (no time-dependent predictors) in the first row of panels, and the time-dependent effects Cox regression model in the second row of panels. Each plot shows the 95% confidence intervals about the respective survival curves. The first column of the figure is the survival curve for all possible combinations of the predictors. The second column of the figure has the survival curve for the set of predictors with values age = 45 years, cigarettes per day = 10, log total cholesterol = ln(300), diabetes status = 1 (patient has diabetes), and sex = 1 (male), which corresponds to Table 7.12. Clearly, the survival curves of the time-dependent Cox regression are different.

The exp(coef) variable for the Cox models is the hazard ratio (HR), which is the multiplicative effect of the predictor on the HR function. A HR = 1 indicates the predictor has no effect on the hazard, HR < 1 indicates a reduction in the hazard, and HR > 1 indicates an increase in the hazard. Absent the proportional hazards assumption, HRs can be highly misleading when the proportional hazard functions are not fixed distances among any two or more hazard curves. This is analogous to a linear regression fit of a data set in which the response and predictors are not linearly related.

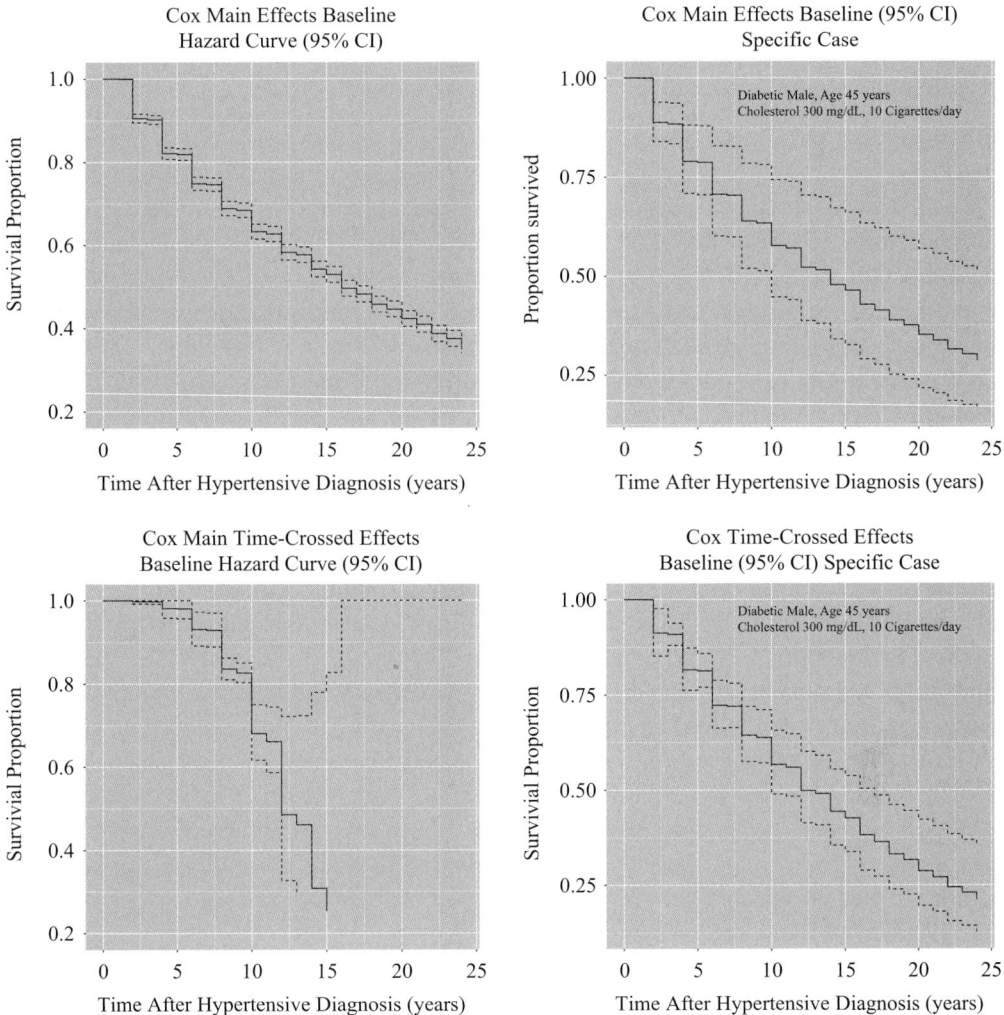

Figure 7.6 Survival curves for the Cox Proportional Hazards model (main effects only) and the time-dependent effects Cox regression model. Each plot shows the 95% confidence intervals about the respective survival curves. Note: the proportion surviving (vertical axis) scales are 0 to 1.

For Cox models beyond the simplest (and a time-interaction makes it nonsimple) we suggest looking at predictions instead of interpreting the coefficients directly. For example, we made predictions for patients with values age = 45 years, cigarettes per day = 10, log total cholesterol = ln(300), diabetes status = 1 (patient has diabetes), and sex = 1 (male), and compared to the same conditions for the CPH. These comparisons, with associated survival function plots make the direction and magnitude of the effects clear. We begin by analyzing the interactions. The significant log total cholesterol interaction term suggests that the effect is different by year. Thus, for example the age term describes the effect of age when gender = 0, or the age effect for males. It is evident that for log total cholesterol the log hazard rate decreases with each year of age by 0.598, and this effect is significant, $p < 0.05$. The age effect, however, is marginally severe for older patients with the effect increasing by

1.037 ($p \ll 0.05$). Hence, the time to a diagnosis of hypertension for older subjects occurs sooner than for younger patients. The time to hypertensive diagnosis increases as the total cholesterol of a patient changes with time.

7.4 Summary

Survival analysis involves non-Gaussian and correlated data. The non-Gaussian aspect arises from the use of survival functions which may be from Gaussian distributions, but more commonly, they are not. This is particularly true in reliability analysis in which failures are modeled with exponential probability functions, lognormal probability functions, and Weibull probability functions, to name a few. Correlation derives from survival depending, of course, on time, and that possible predictor variables of survival often are also time-dependent. We saw that this is the case with the Framingham Heart Study data using time to diagnosis of hypertension with the time-dependent predictors of total cholesterol, age, and the number of cigarettes used per day.

Our time to diagnosis of hypertension survival analysis utilized three common modeling methods: life tables, the Kaplan-Meier method, and two variants of the Cox models. Life tables is a model that gives the probability of a subject not surviving or a unit failing at specified times. They are derived directly from the cumulative probability distribution function. Life tables are constructed using fixed time intervals. The tables give straightforward access into the survival of a particular population. Life tables are a convenient time grouping tool for survival data.

The Kaplan-Meier method does not group patient entry times into intervals as does the life tables. It uses each subject's time of the event of interest to produce survival statistics. For data sets with no censoring, both the Kaplan-Meier analysis and the life table analysis result in the same survival statistics. When censored data are present, the Kaplan-Meier method is used to account for the censored data, thus giving survival statistics that are unbiased relative to the censoring.

The Cox proportional hazards model is a semiparametric modeling approach to describing the relationship of predictors to survival times unlike that for life tables and the Kaplan-Meier method. To use a Cox proportional hazards model, the shape of the hazard proportions must be the same throughout the study period for all observations and all study subject groupings. Once events begin, the rate of occurrence must proceed at the same rate throughout the study for all the study groups. This implies that the survival functions (and the log-survival functions) are roughly proportional throughout the study period.

The Cox proportional hazards model assumes the model predictors are not dependent on time. The Cox time-dependent predictor model, however, makes no assumption on whether the predictors are time dependent. The analysis of the Framingham Heart Study time to hypertension diagnosis showed that the Cox proportional hazards model did not represent the data as well as did the Cox regression model. We must consider carefully whether the predictors are better represented as independent of time, or whether they indeed are dependent on time.

7.5 Further Reading

Hosmer et al. (2008) provide a very understandable introduction to using and interpretation survival analysis models. The guide to survival analysis of Allison (2010) is based on data to introduce the methods of survival analysis, targeted to use with the SAS® statistical analysis system, and requires only a minimal knowledge of the software.

Tabachnick and Fidell (2013) have a very readable chapter on survival and reliability analysis. Three software packages are used on the various example analyses.

The Tobias and Trindade (1995) text on applied reliability has easily followed modern reliability techniques. It is a practical approach for engineers, scientists, and statisticians working with applied reliability problems.

Meeker and Escobar (1998) have extensive treatments of the statistical methods for reliability data analysis. The examples are computed using S-PLUS, which is easily modified to run in R.

Finally, Ebeling (2004) treats maintainability analyses in addition to reliability analysis. An elementary knowledge of statistics is sufficient to use this text.

8

Longitudinal Response Models

8.1 Longitudinal Data

Researchers often collect data over time with the intention of predicting changes, growth, or other effects and associations over time. While there are many techniques for collecting data over time, the most powerful is the collection of panel data. Panel longitudinal data refer to observations made repeatedly on the same individuals or units over time. For example, health researchers may record physiological measures of patients such as systolic blood pressure during monthly visits to each patient's primary care physician; market analysts may record the GDP of a number of countries during each financial quarter of each year over a three-decade span; educational psychologists may track individual students' motivation at multiple times during a school year. Researchers have various motivations for recording panel longitudinal data. There may be an interest in predicting differences, changes, or growth in the outcome of interest across time. There may also be an interest in the effects of other predictors on the response of interest. Researchers may also be interested in changes in the outcome between subjects, in addition to changes within subjects over time.

When modeling longitudinal panel data, it is important to account for the "repeated" nature of the data. Specifically, when responses are recorded multiple times for the same individuals or observational units, those responses are generally associated. This correlation among the outcomes, or autocorrelation, is a violation of the assumption of independence applied to normal regression models. Therefore it is important to construct models that allow for nonindependence of responses.

For example, when recording GDP for a number of countries during each quarter over the span of many years, it is reasonable to assume that successive measures of GDP for the same country will be positively correlated. If these correlated observations are treated as independent, it would appear that a number of different observations are showing similar behavior, thus supporting a certain relationship. However, in truth it is the same country's repeated responses showing this relationship, which is misleading and can result in incorrect reporting of significant associations. Correlated or longitudinal modeling and testing techniques are necessary to account for this potential inflation of significance.

As another example, when recording individual students' motivation over the course of a school year, it might be of interest to pursue modeling and testing techniques that specifically allow for the investigation of trends in student responses over time. In fact, the intention of recording longitudinal data is often to examine such dynamic patterns or relationships over time. Longitudinal modeling techniques allow for these effects to be detected, while cross-sectional models do not.

Data analysts can generally organize longitudinal modeling and testing techniques into three categories: marginal approaches, conditional approaches, and transition models. Marginal approaches involve direct specification of the nature of the correlation among responses by the analyst. Often the source of this autocorrelation is not determined or of interest; the inherent relationships among repeated responses are accounted for to obtain trustworthy estimates of the effects of interest. Conditional approaches, on the other hand, involve identification of sources of random variation, and estimation of such variation in baseline values and effects. After accounting for the effects of individual characteristics, effects of interest can be investigated. Finally, transition models allow the outcomes to depend on previous values of the outcome of interest. The marginal and conditional modeling approaches, along with other relevant statistics, will be discussed throughout the chapter.

8.2 Autocorrelation in Longitudinal Data

It is important to identify and properly account for the inherent correlation among repeated observations on the same individuals or units. If this autocorrelation is not accounted for, marginal or spurious relationships are likely to be identified as significant. When applying longitudinal data methods to panel data, researchers are often interested in trends of outcomes over time, in the variation of responses within and between subjects, in identifying predictors that affect trends or subject-related variation, and in identifying relationships between predictors and outcomes as with cross-sectional studies.

8.2.1 Descriptive Analysis

It is always important to evaluate data descriptively prior to building models and making inferential decisions. Descriptive and exploratory analyses of data have been discussed earlier in the text; however, longitudinal data include some exploratory approaches that differ from those generally applied to cross-sectional data that do not include observations made over time.

8.2.2 Scatter plots

Appropriately chosen plots can provide useful information in model selection and also in framing conclusions within the context of the data application. When analyzing longitudinal panel data, it is common to investigate scatter plots of the response versus an appropriate measure of time, often referred to as a "time plot." Such plots can give a quick indication of obvious trends in responses over time.

The time plot is often augmented with a "spaghetti plot," in which response values over time for the same subjects are connected by line segments. Thus it may appear that a rough "string" connects each subject's responses in the plot across time. In this way it is easy to identify highly unusual subject patterns and extreme responses. It can also be helpful to consult a spaghetti plot to identify whether subjects themselves break into groups; for example, one group of subjects may show a general increase over time while another group of subjects may show a general decrease over time. This is challenging to identify without connecting responses from the same subjects using line segments in the plot. When a large number of subjects is represented in a single data set, we often create spaghetti plots in which a random selection of subjects have responses connected by lines. Alternatively, specific percentiles of subjects can have responses connected by lines (for example, using baseline values to identify percentiles, the 10th, 50th, and 90th percentile individuals will have responses connected by lines while all other subjects' responses remain unconnected across times).

When the data allow, certain interaction plots can provide useful information. Recall that within traditional factorial ANOVA analyses the interaction plot is used to graph cross-classified response averages across one factor, with responses connected according to the levels of a second factor. This allows the analyst to graphically examine whether the trend or pattern in the average responses across one factor changes according to the levels of another factor. Parallel lines indicate similar patterns, while nonparallel lines indicate changes in the trends. Within the context of longitudinal data, the response averages are typically plotted against time on the horizontal axis; any second factor can be used to connect averages to produce an interaction plot. This allows the analyst to determine whether trends over time change according to any independent factors of interest, such as gender or age group. For these interaction plots to be feasible, it is necessary to have grouping or classification predictors that do not change over time.

When time-changing predictors are present in the analysis, it can be useful to create predictor change plots in which the difference scores in the response $(y_{iT} - y_{i1})$ are plotted against the same scores for a single predictor, $(x_{iT} - x_{i1})$. Here $(y_{iT} - y_{i1})$ represents the growth in the response from time 1 to time T, and $(x_{iT} - x_{i1})$ represents the same growth for the predictor x. This allows the data analyst to roughly determine whether the growth in the outcome is associated with the change in any predictor.

8.2.3 *Autocorrelation Plots*

In addition to evaluating trends over time, some exploratory longitudinal plots can be used to assess the nature of the autocorrelation in the data. Using an otherwise appropriate linear model that does *not* account for the autocorrelation in the data, residuals can be obtained for each individual at each time. A correlation analysis of these residuals can provide insight into the nature of the relationships among responses across times. For example, a scatter plot matrix of residuals across different times will show the strength of the relationships between residuals at any two times of observation. These plots can be supported with a correlation matrix using residuals at different times.

The autocorrelation function, common for time-series analyses, can also be applied to residuals at different times. Aggregating all of the residuals associated with observations separated by a single time, an estimate of the correlation in the data across a single time can be estimated. This is the autocorrelation associated with a "time lag" of 1. Similarly, all of the residuals associated with observations separated by two times can be aggregated, and an estimate of the correlation in the data across two times can be estimated for a time lag of 2. In this way a table of correlation values associated with different time lags can be constructed, and an associated autocorrelation plot can show the changes in these estimated autocorrelations.

The autocorrelation plot depends on some strong assumptions about the data. For example, when the data are aggregated to calculate correlation for all residuals separated "by a single time," implicit is the assumption that all time changes are equivalent. For example, this aggregation is appropriate when all observations are taken at equally spaced times, such as GDP records taken at the end of each quarter. In this case all observation times are separated by three months, and differences between any two consecutive measurements will always represent three months of time. On the other hand, the aggregation involved in the autocorrelation function will not be appropriate for unequally spaced times of observation. Supposing measures of individual students' motivation are taken in August at the beginning of the school year, again in December, in January, and in May, the observations are not equally spaced. Consecutive measurements might represent one month or four months, which is not captured by the autocorrelation function.

A stronger assumption implicit in the construction of the autocorrelation function is that of *stationarity*. Briefly, a trend or process over time is stationary if the relationships across times depend only on the time change, but not on the time from which the initial reference is made. For example, if the relationship between successive measures of systolic blood pressure is stronger for the first two measures of the study than for the final two measures of the study, the process describing trends in systolic blood pressure is not stationary because the nature of the relationship depends on which specific times are under consideration. Most longitudinal data are assumed to follow stationary processes, and such an assumption will be made for all data analyses in this chapter.

8.2.4 Variograms

The variogram provides a stronger visual tool for evaluating the autocorrelation in longitudinal data than the autocorrelation function, and does not require assumptions of equally-spaced times or stationarity of processes. The variogram involves plotting a measure of variation on the vertical axis, versus a measure of actual differences in times of observation on the horizontal axis. The measure of variation on the vertical can be thought of as a measure of variation that is not accounted for by the association among outcomes over time. In most cases (and in particular for stationary processes) it is proportional to 1 minus the autocorrelation values, or the complement of the autocorrelation, scaled by the variation in the data. Small values in the variogram suggest strong association between times; larger values suggest less variation. Often the variogram includes a smooth curve (such as a loess curve) to help identify trends within the plot.

Some shapes commonly arise in variograms of longitudinal data. A variogram that begins with very small values that increase steadily as the time lag increases suggests that the autocorrelation is initially strong but decays over time. This could be considered an autoregressive process, in which autocorrelation is assumed to decay with increased time lag. A variogram that shows little change in the values with time lag would suggest a compound symmetry (or exchangeable) process, in which the autocorrelation is constant across all possible time lags. A variogram that begins with very small values but immediately becomes very large suggests that initial autocorrelation disappears as soon as outcomes are separated by any kind of time lag. This could indicate a truncated autocorrelation structure, in which correlation exists between responses that are consecutive in time but not between any responses separated by more than a single time of observation.

8.2.5 *Modeling Longitudinal Data*

Longitudinal data models generally fall into one of three categories: marginal models, conditional models, and transition models. The modeling options differ in terms of the decisions the analyst must make in constructing the model, the associated assumptions and estimation methods, and the resulting conclusions that can readily be made.

Marginal longitudinal models include terms that directly describe the correlation among responses. With marginal modeling approaches, the goal of the researcher is to make population-averaged interpretations. Population-averaged interpretations are also made using ordinary regression models: the coefficients represent an expected difference in the mean response between two populations that differ, on average, according to one of the predictors in the model. Within the context of longitudinal panel data, marginal models allow us to evaluate the effects of predictors while accounting for the autocorrelation in the data. For example, an educational psychologist would be able to use a marginal modeling approach to evaluate the impact of a specific teaching style on students' motivation, while accounting for the changes in student motivation over the course of a school year. In this case there may not be a specific interest in changes over time, but the longitudinal nature of the data must still be accounted for by specifying the nature of the autocorrelation among responses within the model.

Conditional longitudinal models include terms that describe individual tendencies of subjects or observational units in the study. With conditional modeling approaches, the goal of the researcher is to make subject-specific interpretations. Using a subject-specific interpretation, coefficients represent an expected difference in the mean response as an individual changes over time. For example, health researchers would be able to use conditional modeling to predict the expected change in an individual's systolic blood pressure over time as the dosage of blood-thinning medication is changed. In this case there is a deliberate interest in individual changes over time, according to specific predictors.

Transition longitudinal models include terms that represent the direct effect of previous outcome measures on future outcome measures. Transition models will not be considered further in this chapter.

8.3 Marginal Models

All traditional normal regression models can be considered marginal models, in the sense that effects are averaged across populations and comparisons are made between populations. Likewise, marginal longitudinal models involve comparisons between populations, and appropriate conclusions from such model must be worded as population-averaged effects. For example, in the hypothetical situation of collecting systolic blood pressure over time, the analyst might be interested in the differences in mean systolic blood pressure between males and females, but not necessarily in individual trends in blood pressure over time. Given the repeated recording of systolic blood pressure of individuals over time, it is necessary to account for the inherent autocorrelation in the data when making comparisons between groups. A marginal longitudinal model could be applied, thus accounting for the nonindependence of responses over time, but allowing for a population-averaged conclusion such as general differences between males and females.

8.3.1 Generalized Estimating Equations

The most popular method of modeling longitudinal data to produce population-averaged conclusions is the generalized estimating equations (GEE) approach. Briefly, the GEE approach involves specifying two components to construct a marginal longitudinal regression model: a model for the mean of the response (as is typically specified in all normal regression models) along with some description of the expected structure of the autocorrelation in the longitudinal responses. For example, in modeling repeated measures of student achievement as the outcome variable, researchers might first specify a model for the mean by constructing what looks to be a normal regression model with a measure of student achievement as the outcome, and including any predictors of interest as is typically done with a regression model. The second step of the GEE process is to specify the structure of the autocorrelation, for which researchers might specify that the associations among repeated measures of student achievement for the same individual diminish as the observations become further apart in time. Notice that it is not necessary to specify values for the autocorrelation; it is only necessary to describe the structure of the autocorrelation.

The first component of the marginal model, which represents the mean model, can be written generally to include all of the models discussed in this text:

$$g(\mu_{it}) = \beta_0 + \sum_{j=1}^{J} \beta_j x_{itj}, \tag{8.1}$$

where μ_{it} is the mean of the outcome of interest, for subject i at time t. The x_{itj} represent observed predictor values for subject i at time t, and the β_j are the unknown regression coefficients, as with normal regression. The primary interest is to estimate these β_j in order to assess the impacts of the independent variables and to make predictions about the mean, μ_{it}. The notation $g()$ is used to indicate the transformation that relates the mean of the response data to the predictors and parameters. For logistic regression applied to binary responses, this will be the logit transformation. For normal regression applied to continuous responses, this will be the identity transformation (i.e., no transformation). Notice the structure, and also the interpretations of parameters, are both very similar to those of models encountered previously in the text. The second component of the marginal model, which represents the nature of the associations among outcomes of interest, is determined by the analyst's choice of working correlation structure.

8.3.2 Working Correlation Structure

The component of the marginal model that represents the expected structure of the autocorrelation is chosen by the data analyst to reflect the expected relationships among the repeated responses for subjects. Using existing literature, pilot data, a training portion of the final data set, or simply intuition about the outcome of interest, the analyst must decide on the most likely nature of the associations among responses. This is related to the response covariance structure, but is more easily understood through the response correlation structure, which is related to covariance through a direct transformation. Ultimately, the analyst chooses what is referred to as a "working correlation structure," which is an outline of the structure of the autocorrelation for each subject. For example, consider the case of recording three comparable measures of student achievement throughout the academic year, such as standardized tests given in September, January, and May of the academic year. An analyst might propose that these three outcomes for any individual student are all equally related; that is, the first and second test scores are correlated just as much or as little as the second and third test scores (and also the first and third test scores). This is known as the "exchangeable" or the "compound symmetry" working correlation structure. This is often represented in matrix form, in which each entry represents the value of the autocorrelation between responses at different times, for the same individual:

$$\mathbf{R}(\rho) = \begin{bmatrix} 1 & \rho & \rho \\ \rho & 1 & \rho \\ \rho & \rho & 1 \end{bmatrix}.$$

Here the first row corresponds to the first time, the second row to the second time, and the third row to the third time. Similarly, the columns are associated with the first, second, and third times. The value in the first row, third column, for example, represents the analyst's expectation for the autocorrelation between outcomes at the first and third times. Using this notation, the 1's indicate that the correlation should be perfect between responses at the same time; for an individual, this is simply stating that a value at one time is perfectly correlated with itself. The ρ's indicate the value of the autocorrelation between responses at different times. These values are all identical, because under the exchangeable structure all responses are correlated equally. Notice the actual *value* of this correlation need not be included in the working correlation structure. Instead the numeric value of ρ will be estimated through GEE, using the data.

  A researcher can propose a number of alternative working correlation structures. It may be more appropriate to assume that the correlation between outcomes is strongest for consecutive responses, but weakens over time. That is, the scores for September and January will have a stronger correlation than will the scores for September and May. As more time passes, the correlation weakens. This is known as an "autoregressive" working correlation structure, and can be expressed as

$$\mathbf{R}(\rho) = \begin{bmatrix} 1 & \rho & \rho^2 \\ \rho & 1 & \rho \\ \rho^2 & \rho & 1 \end{bmatrix}.$$

Specifically, the working correlation structure shown is the "first order" autoregressive structure, because the power of the correlation parameter increases by 1 for each additional step of time, reducing the value of correlation. Two more common working correlation structures are the unstructured and independent working correlation structures. Under the unstructured assumption, all values of autocorrelation can differ from each other:

$$\mathbf{R}(\rho_{12}, \rho_{13}, \rho_{23}) = \begin{bmatrix} 1 & \rho_{12} & \rho_{13} \\ \rho_{12} & 1 & \rho_{23} \\ \rho_{13} & \rho_{23} & 1 \end{bmatrix}.$$

Notice the unstructured assumption requires the estimation of three correlation parameters instead of one. Under the independent working correlation structure, there is no association between repeated observations of the same individuals:

$$\mathbf{R} = \begin{bmatrix} 1 & 0 & 0 \\ 0 & 1 & 0 \\ 0 & 0 & 1 \end{bmatrix}.$$

An analyst can propose any working correlation structure that legitimately describes correlation, beyond those four described above. The four discussed will be available in most software packages.

Typically, the correlation parameters from the assumed working correlation structure, such as ρ, are treated as "nuisance" parameters. The values are estimated as part of the process of fitting the GEE model, but are often not reported or evaluated for significance. The focus of the marginal analysis is typically on the β_j from the mean model, but by selecting and estimating the parameters of a working correlation structure, the results of the mean model can be interpreted as "accounting for" the repeated observation of individuals. The parameters β_j then have population-averaged interpretations. That is, accounting for the repeated observation of subjects, β_j represents the expected difference between populations that differ, on average, by one unit of the predictor x_{itj}.

8.3.3 *Marginal Model Fit*

Marginal longitudinal regression model fit is commonly assessed using a combination of residuals and global fit values. For most analyses, data analysts use measures of fit to compare marginal modeling options to determine the most appropriate choice. For example, an analyst may compare fit measures among the choices of autoregressive, exchangeable, and unstructured working correlation structures. In this way fit is evaluated in terms of identifying the most appropriate model from a list of options, but for GEE models fit is rarely assessed in a standalone sense.

Many types of residuals are available for the GEE estimation method, including raw residuals, Pearson residuals, and deviance-type residuals. (Because no full likelihood function is used for GEE estimation, true deviance cannot be calculated. However, using the mean-variance relationship and the implied type of response, deviance formulas from comparable likelihood-based estimation can be used to calculate deviance-type residuals.) As with all models, extreme values of residuals indicate poor fit of the model to the data.

Information-type criteria have been developed for the GEE method of estimation to serve a role similar to the AIC and BIC of likelihood-based estimation methods. The quasi-likelihood under the independence model criterion (QIC) is the most common such statistic, and can be used to identify the most appropriate working correlation structure. QIC can be interpreted similarly to AIC, as a measure of information lost by a model, such that smaller values are preferred. Assuming a working correlation structure has been selected appropriately, an adjustment to QIC referred to as QICu is often used for variable selection.

8.4 Conditional Models

The most general approach to modeling longitudinal panel data is the conditional approach, in which a separate error structure is constructed for the random subjects or individuals who are observed or measured repeatedly. This modeling approach can also be referred to as mixed models, random-effects models, and often as hierarchical or multilevel models. When using this approach, the researcher is implicitly assuming that there are individual-specific factors that cause variation in the outcomes between subjects. It is this variation that is captured by the additional randomly varying terms that are included in conditional models. The resulting parameters can be interpreted using subject-specific conclusions, in which estimates apply to the expected change in the response for an individual over time.

There is no universally preferred estimation method for conditional longitudinal regression models, as with the GEE approach for marginal modeling of panel data. Conditional models can be differentiated according to the number of additional random terms included, and the nature of the inclusion of those random terms. In the following discussion there will be a focus on random-intercept models, and on random-slopes models.

8.4.1 *Random-Intercept Models*

The random-intercept model is the most basic of conditional longitudinal models. A single error structure is introduced to account for the fluctuation of individual outcomes across the population of subjects. Essentially each subject is allowed a random "baseline" value of the expected outcome, from which the effects of individual predictors can be added to make predictions of the response. Because there is only one additional randomly varying term and it only affects the baseline value (and not the effects of predictors), the model can be viewed as allowing the intercept or constant in the regression model to vary randomly. Thus this model is referred to as a random-intercept model.

The random-intercept model is potentially more straightforward to select than the common marginal GEE model. The analyst need only choose to use this model, implicitly identifying the individuals of the panel data collection as an additional source of random variation, then specify the model for the (conditional) mean of the response of interest. The model can be written similarly to other regression models,

$$g(\mu_{it}) = \beta_0 + \sum_{j=1}^{J} \beta_j x_{itj} + u_i, \tag{8.2}$$

where the x_{itj} represent observed predictors' values, and the β_j and u_i are unknown effects to be estimated through the model. The parameter u_i represents the random fluctuation in the baseline value for each individual, and is intended to capture discrepancies in individual tendencies across subjects. It is typically assumed that this random variance term is normal and can be included additively in the regression model, although both of these assumptions can be changed. The parameter u_i will have a variance associated with it, but this term is estimated as a "nuisance" parameter much like the correlation terms in the GEE approach. For example, consider the case of recording patient systolic blood pressure on three consecutive monthly visits. In applying a random-intercept model to predict expected blood pressure, the random effect u_i would be intended to account for random variation in individual tendencies that causes individuals to have different baseline levels of blood pressure, regardless of the other predictors used in the model. These individual tendencies are then accounted for by including this term in the model.

Most of the other terms in Eq. (8.2) remain as with the marginal longitudinal regression model. The x_{itj} represent predictor values for subject i at time t, the β_j are the regression coefficients, and $g()$ is used to indicate the transformation that relates the mean of the response data to the predictors and parameters of interest. In this case μ_{it} represents the "conditional" mean of the response, conditioned on the random fluctuations u_i. This means μ_{it} represents what we expect the outcome to be for a given individual, with the effect of u_i included.

There is no specification of any working correlation structure for conditional models. The inclusion of the random effect u_i imposes a certain structure on the autocorrelation among responses. For normal models with an additive u_i, the exchangeable correlation structure results from the model. But the analyst need not select this aspect of the model; it is only necessary to specify the mean model and determine that a conditional approach will be taken.

The parameters β_j associated with the predictors are of primary interest, and have subject-specific interpretations. That is, given an individual with a specific random baseline that takes into account the effects of u_i, β_j represents the expected change associated with an increase in the predictor x_{itj} for that individual. In other words, the subject-specific interpretations from a conditional model allow for the types of interpretations of changes over time that are often desired from collecting longitudinal panel data. For example, in the case of modeling systolic blood pressure over time using a conditional longitudinal model, the coefficient for the predictor of "cholesterol" would indicate the expected impact on an individual's blood pressure when cholesterol increases over time.

8.4.2 Random-Slopes Models

The random-intercept model from Eq. (8.2) can be written in equation form to showcase the view of the random fluctuations u_i as adjustments to the intercept

$$g(\mu_{it}) = (\beta_0 + u_i) + \sum_{j=1}^{J} \beta_j x_{itj}. \tag{8.3}$$

In this presentation the term u_i is treated as a random error associated with the intercept term, consistent with the interpretations discussed in the previous section. As an extension, the random-slopes conditional longitudinal model allows for all of the β_j in the model to have associated random errors, including the slopes associated with each predictor,

$$g(\mu_{it}) = (\beta_0 + u_{0i}) + \sum_{j=1}^{J} (\beta_j + u_{ji}) x_{itj}. \tag{8.4}$$

In the random-slopes model, there are random error terms u_{ji} associated with all model predictors as well as the intercept. If it is not desired to have random fluctuations for all slopes, this presentation of the model can be adjusted to allow any combination of model predictors to have randomly varying slopes. The interpretation of the terms u_{ji} is similar to that of the original random effect u_i from the random-intercept model. Each u_{ji} can be thought of as random variations of the *effect* of each predictor on the mean response, among the subjects in the study. For example, in the case of modeling student achievement over time using a random-slopes model, the random error term u_{ji} associated with the prior grades predictor would represent random variation in the impact of prior grades on achievement across the population of students considered for the study. However, the random effects u_{ji} are typically treated as nuisance parameters: the variation associated with each must be estimated and accounted for, but the values are often not reported and hypothesis tests associated with the u_{ji} are rarely of interest.

The primary interest typically remains the slopes β_j. All of the standard regression terms β_j can be thought of as the expected effect of a predictor, after taking into consideration the tendencies of the individual (as represented by all of the u_{ji} terms). Interpretations of the β_j can be made similarly to those of the random-intercept model. That is, after accounting for the effects of the individual tendencies of subjects in the study, each β_j represents the expected change in the mean response when an individual's value of predictor x_{itj} increases over time.

8.4.3 Conditional Model Fit

Fit of conditional longitudinal models is assessed using a combination of residual analyses and global fit statistics. The familiar information criteria, such as AIC and BIC, can be applied for conditional model comparisons. Because estimation is likelihood-based, we also have available the global deviance value. The model deviance should be distributed approximately as a χ^2 statistic, and therefore the deviance divided by its degrees of freedom should be close to 1. If this ratio greatly exceeds 1, the model may fit the data poorly.

There are a number of residuals available for conditional longitudinal models. Conditional model residuals are identified by two properties: the treatment of the random effects in the model, and the scale on which the residuals are calculated. Conditional residuals include predictions of the values for sources of random variation in the model, while marginal residuals use the average value of zero for all sources of random variation. Mean-scale residuals involve raw differences between observed values and predicted means, while linearized-scale residuals involve transformed data and direct values of predictors. We will prefer to use conditional Pearson and deviance residuals, which are calculated on the mean scale. The Pearson residuals can be written as follows:

$$r_{\mu,it} = \frac{y_{it} - \hat{y}_{it}}{\hat{\text{Var}}(y_{it}|u_i)}, \tag{8.5}$$

where y_{it} is the observed outcome for subject i at time t, $\hat{y}_{it}$ is the associated predicted value, including estimated values for the random sources of variation, and $\hat{\text{Var}}(y_{it}|u_i)$ is an estimate of the variation of the raw residual in the numerator. Conditional Pearson residuals resemble the residual calculations for normal linear models, and can be used to evaluate assumptions of normality, constant variation (if applicable), and general appropriateness of the model. Deviance residuals, as described in Chapter 2, can also be calculated according to the assumed distribution of the response and used similarly to Pearson residuals.

8.5 FHS Analysis: Probability of Hypertension

We are interested in modeling the probability of individuals showing evidence of hypertension using the Framingham Heart Study data. Specifically, we would like to predict the probability of hypertension over the three waves of data collection, using the following variables as predictors.

- Sex: sex, male (1) or female (2)
- Diabetes: the presence of diabetes
- Age: patient age in years
- Cigarettes: the number of cigarettes smoked per day
- Cholesterol: the total serum cholesterol at each follow-up

In particular, we would like to describe differences among populations such as males and females, diabetics and non-diabetics, and individuals in different age groups. Due to the panel nature of data collection, we know we must account for the effect of repeated observation of the same individuals.

In this case an appropriate model is a marginal longitudinal logistic regression model. We will apply a marginal model because of the interest in describing discrepancies among populations and of accounting for the autocorrelation associated with repeated observation of subjects, but not in the development of individuals over time. We will begin by presenting some exploratory analyses, followed by an investigation into the autocorrelation in the data. Then we will present an appropriate marginal longitudinal logistic regression model, we will evaluate the fit of the model, and then proceed to interpret parameter estimates and make predictions using the model.

8.5.1 Exploratory Data Analysis

Some exploratory analyses of these data were presented in Chapter 1, but due to the binary nature of the data further investigation is required. We first construct a contingency table showing cross-classified counts of prevalent hypertension versus categorical independent variables, across times of observation.

The values in Table 8.1 show that the odds of hypertension increases with time, because the raw observed odds of hypertension during the first period of observation ($1\,430/3\,004 \approx 0.476$) is less than the same raw odds observed during the second period of observation ($1\,959/1\,972 \approx 0.993$), which is less again than the odds associated with the third period of observation ($1\,955/1\,308 \approx 1.495$). Table 8.1 also shows that the odds of hypertension during the first period of observation is lower for females ($799/1\,691 \approx 0.473$) than it is for males ($631/1\,313 \approx 0.481$), and that these odds both increase with each successive period of observation. Similarly, the odds of hypertension for diabetics during the first period of observation ($67/54 \approx 1.241$) is greater than that of non-diabetics ($1\,363/2\,950 \approx 0.462$).

Table 8.1 *Framingham Heart Study observed counts of hypertension by categorical predictors.*

		Time of data collection					
		Period 1		Period 2		Period 3	
Hypertension		Yes	No	Yes	No	Yes	No
Total		1 430	3 004	1 959	1 972	1 955	1 308
Diabetes	Yes	67	54	116	39	202	52
	No	1 363	2 950	1 843	1 932	1 753	1 256
Sex	Male	631	1 313	850	841	845	542
	Female	799	1 691	1 109	1 130	1 110	766

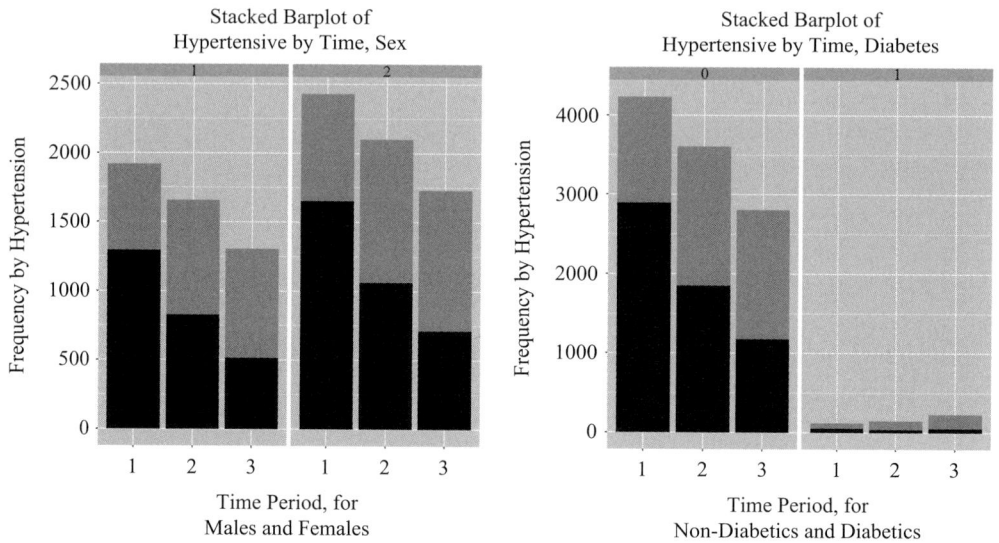

Figure 8.1 Framingham Heart Study stacked bar plot by study period, light shading for hypertensive and dark shading for not hypertensive, males and females (left panel), non-diabetics and diabetics (right panel).

Both the odds of hypertension for diabetics and for non-diabetics show an increase with periods of observation.

The observations from Table 8.1 can be visualized using stacked bar plots of hypertension over time. Figure 8.1 shows stacked bar plots of hypertension counts over time, split by sex in the left panel, and split by diabetes in the right panel. Using these plots, the odds are visualized as a comparison of the area associated with hypertension to area associated with no hypertension. The plot in the left panel of Figure 8.1 shows that these relative areas of light versus dark generally increase in favor of hypertension, suggesting increased prevalence of hypertension over period of observation for both males and females. The plot also shows the attrition effect, as the raw numbers of males and females also decrease with observation period. The plot in the right panel of Figure 8.1 shows the same dynamic for diabetics and non-diabetics: the proportion of plot area in favor of hypertension increases with period of observation. Notice the raw number of diabetics increases with observation period.

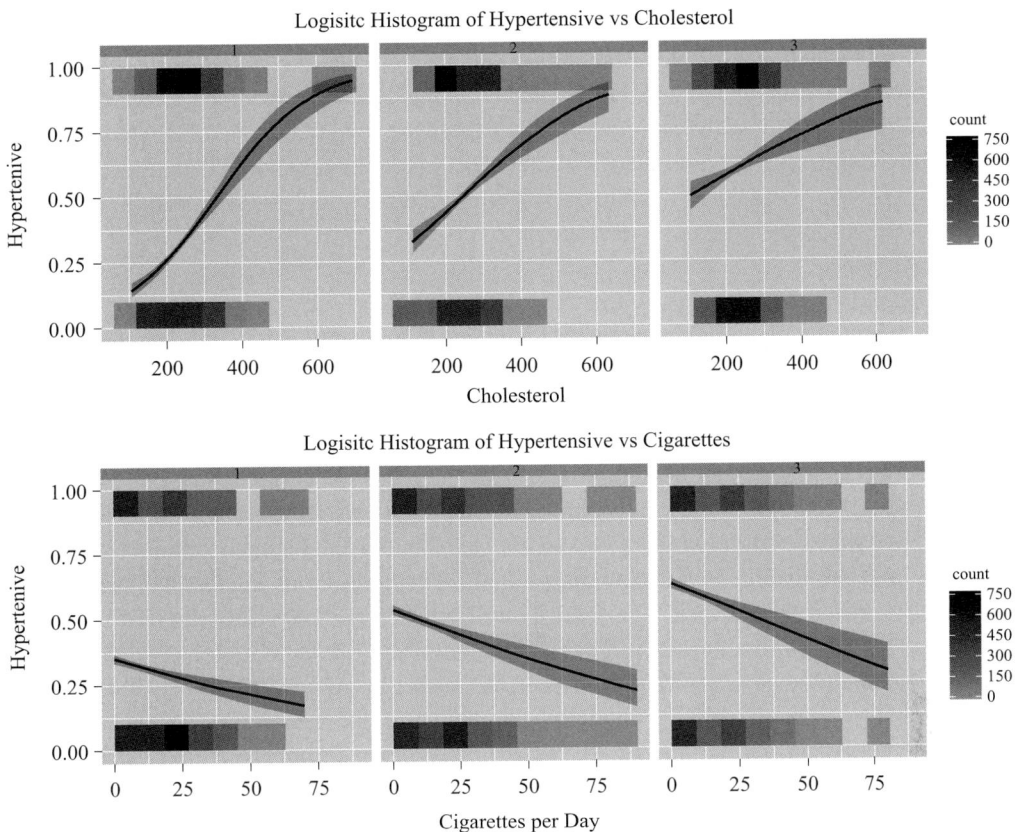

Figure 8.2 Framingham Heart Study logistic histogram plot of hypertension by total cholesterol, for each study period (left panel), logistic histogram plot of hypertension by cigarettes smoked per day, for each study period (right panel), darker shading indicates greater numbers hypertensive.

Continuous predictors can be investigated using logistic histogram plots, which indicate the frequency of observations of outcome values across values of predictors. For longitudinal analyses, it is informative to view such plots at each time of observation. The plots in Figure 8.2 show such plots for the predictors total cholesterol in the panels of the top row, and cigarettes per day in the panels of the bottom row.

In the logistic histogram plots for hypertension versus cholesterol, shown in the panels of the top row of Figure 8.2, the smooth logistic curve suggests a positive relationship between cholesterol and hypertension for all three times of observation. This type of relationship makes sense intuitively, and it is reassuring to see the same pattern across study periods. Also notice that this relationship suggests consistently higher likelihood of hypertension for the third study period, when subjects are oldest, because the smooth curve generally shows larger values for the third study period than for the first two.

The plots in the panels of the bottom row of Figure 8.2 shows logistic histogram plots for the predictor cigarettes per day. Notice the smooth logistic curve suggests a negative relationship between cigarette use and hypertension for all three times of observation.

While the relationship is consistent over study periods, the expected chance of hypertension appears to increase over time, corresponding to aging of subjects in the study, because the smooth curves generally show greater values as the study period increases. The trends suggested in all of these exploratory plots should be taken descriptively, without any suggestion of significance, and also without accounting for any of the other predictors in the proposed model. The plots should be taken as indications of possible trends and effects to be seen in a more comprehensive model.

8.5.2 Marginal Longitudinal Model

Because the outcome of interest, hypertension, is a binary variable, we have chosen to apply an ordinary binary logistic regression model. In order to answer the original questions about differences between populations while accounting for the longitudinal nature of the data, we have selected a marginal longitudinal logistic regression model. Such a model allows for conclusions comparing the probability of hypertension or the odds of hypertension between populations, while properly accounting for the expected autocorrelation in repeated measures of hypertension in the same individuals. The continuous predictors (age, number of cigarettes per day, and total cholesterol) will be included as with standard regression models, and the categorical predictors will be included using indicators for sex and diabetes. In addition, the model will include interactions between sex and diabetes, and also sex and cigarettes per day, which will allow evaluation of the differences in the effects of both diabetes and cigarette use between males and females. Finally, the model will include indicators for the second and third study periods, which will control for differences in prevalence of hypertension across times of observation.

Recall that marginal models can be specified in two steps: first a regression model is proposed, and second a structure for the autocorrelation is proposed. The regression model can be written in equation form as follows:

$$\ln\left(\frac{\pi_{\text{Hyp}}}{1 - \pi_{\text{Hyp}}}\right) = \beta_0 + \beta_1(\text{diabetes}) + \beta_2(\text{sex}) + \beta_3(\text{age})$$
$$+ \beta_4(\text{cigarettes}) + \beta_5(\text{cholesterol}) + \beta_6(\text{diabetes} \times \text{sex}) \quad (8.6)$$
$$+ \beta_7(\text{sex} \times \text{cigarettes}) + \beta_8(\text{period 2}) + \beta_9(\text{period 3}),$$

where π_{Hyp} indicates the probability of hypertension and all variables are abbreviated as with the Framingham Heart Study data. Recall that this equation is equivalent to writing the probability of hypertension in terms of all predictors, using a nonlinear relationship,

$$\pi_{\text{Hyp}} = \frac{\exp(\beta_0 + \beta_1(\text{diabetes}) + \cdots + \beta_9(\text{period 3}))}{1 + \exp(\beta_0 + \beta_1(\text{diabetes}) + \cdots + \beta_9(\text{period 3}))}.$$

8.5.3 Examining the Autocorrelation

Before proposing an autocorrelation structure or evaluating the fit of the marginal longitudinal logistic regression model, it is important to descriptively explore the nature of the associations among the responses in the data. Using a nonlongitudinal logistic regression model, omitting the time indicators and the autocorrelation structure, we can use residuals

Histograms, Scatter Plots, and Pairwise Correlations of Residuals

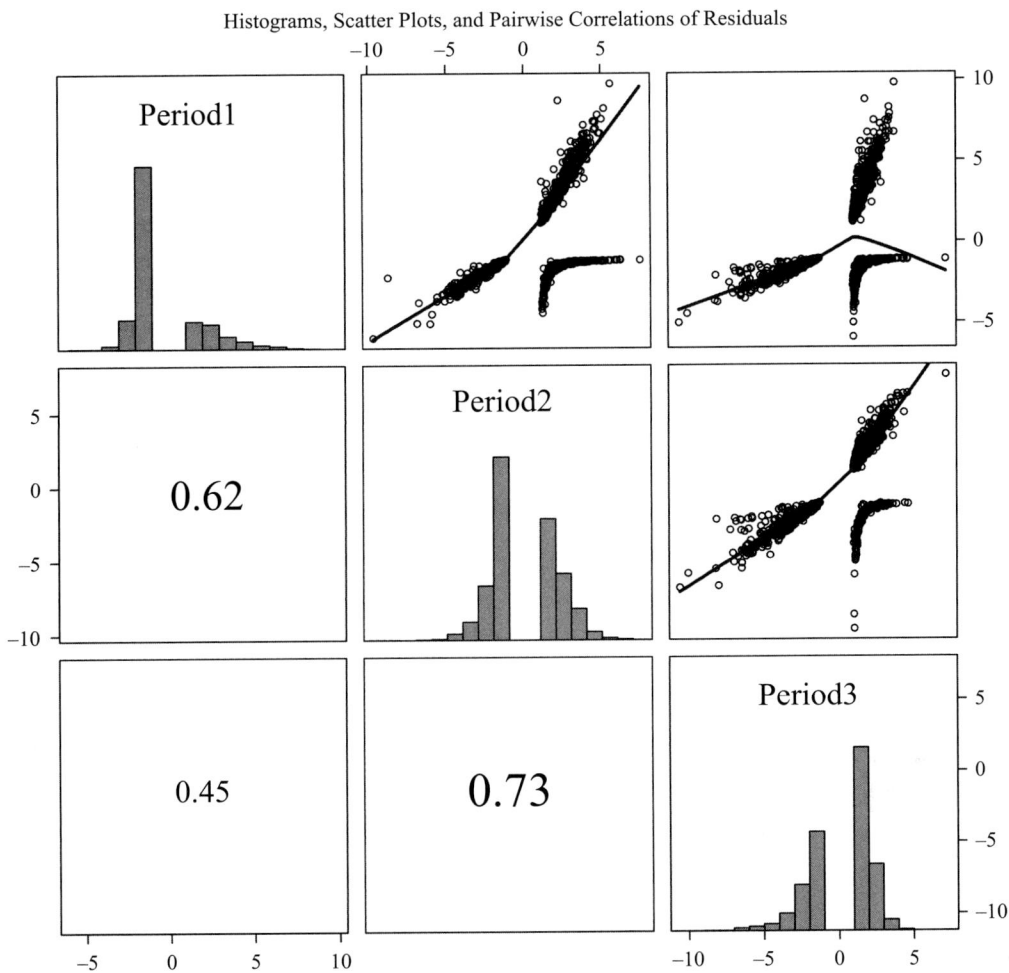

Figure 8.3 Scatter plot matrix of School Survey on Crime and Safety residuals by study period, including histograms on the diagonal, pairwise Pearson correlations, and smooth loess curves.

to get an indication of the nature of the associations among the probability of hypertension over study periods. Figure 8.3 shows a scatter plot matrix of residuals obtained from the three different periods of the study. Recall that these scatter plots do not represent responses or predictors, but rather the remaining unexplained response variation from an initial model.

The scatter plots are used to get a general sense of whether the associations among residuals change with the lag in times. Based on these scatter plots and the superimposed smooth fits, it appears the association among residuals is positive, and strongest between residuals for the second and third observation times but weaker between other study periods.

This conclusion can be supported by considering simple estimates of correlation between residuals at different times, as shown opposite the scatter plots in Figure 8.3. Because the correlation between residuals at consecutive times (0.616 between the first and second times, 0.733 between the second and third times) is larger than the correlation between residuals with a time lag of two (0.453 between the first and third times), it appears

Table 8.2 *Framingham Heart Study QIC values and ranges of marginal deviance residuals for GEE fit using independent, compound symmetry, autoregressive, and unstructured working correlation structures.*

Model	QIC	Residual minimum	Residual maximum	Range
Independent	13 591.86	−2.205	2.181	4.386
Compound symmetry	13 621.74	−2.113	2.168	4.281
Autoregressive (1)	13 623.46	−2.113	2.159	4.272
Unstructured	13 633.65	−2.117	2.157	4.274

the autocorrelation is not consistent across time lags. This decay in autocorrelation suggests an autoregressive working correlation structure may be appropriate.

A variogram could be considered next to further investigate the autocorrelation structure. However, a variogram plots as the horizontal the "time lags" between observations, or the differences between moments of observation. In this case there are only two possible time lags (a lag of 1 between observation times 1 and 2, and observation times 2 and 3; a lag of 2 between observation times 1 and 3). With only two possible lags, it would be difficult to estimate the trend in variation across time lags. The residual correlation analysis is more informative in this case, and we use this residual analysis to propose an autoregressive type of autocorrelation structure.

8.5.4 *Marginal Longitudinal Model Fit*

We assess the fit of the marginal longitudinal model using marginal residuals and summary measures, such as QIC and QICu. Table 8.2 shows the QIC values for GEE models fit using Eq. (8.6) as the regression model and various common working correlation structures.

Based solely on QIC values, the model that loses the least amount of information as compared to the data is the model using the independent working correlation structure, as Table 8.2 shows this model to have the smallest QIC. However, the residual correlation structure showed something resembling autoregressive correlation. Further evaluation of the working correlation options was made using model residuals. Table 8.2 displays the range of marginal deviance residuals for each model, which shows the GEE model with autoregressive (1) working correlation structure to have the residuals with smallest absolute value. Given all of the evidence regarding fit, we proceed to interpret results from the model with the autoregressive working correlation structure.

8.5.5 *Model Parameter Interpretations*

The results of fitting the marginal longitudinal logistic regression model using the autoregressive working correlation structure can be found in Table 8.3, with odds ratios and associated confidence intervals in Table 8.4. The final column of Table 8.3 shows p-values associated with Wald tests for significance. Results showed the effects of both period 2 and period 3 of data collection to be significant. In other words, the data showed evidence of a

Table 8.3 *Framingham Heart Study results of marginal longitudinal logistic regression of hypertension, fit using the generalized estimating equations with autoregressive working correlation.*

Coefficient	Estimate	Std. error	z-value	$Pr(>|z|)$
Intercept	−5.168	0.210	603.29	< 0.001
Diabetes	0.174	0.136	1.63	0.201
Sex (female)	−0.080	0.069	1.36	0.243
Age	0.077	0.004	461.77	< 0.001
Cigarettes	−0.001	0.002	0.30	0.582
Cholesterol	0.003	0.0005	28.61	< 0.001
Period 2	0.403	0.034	139.43	< 0.001
Period 3	0.503	0.051	97.56	< 0.001
Diabetes × sex (female)	0.333	0.191	3.04	0.081
Sex (female) × cigarettes	−0.007	0.004	3.01	0.083

Table 8.4 *Framingham Heart Study estimated odds ratios for hypertension, with 95% confidence intervals.*

Coefficient	Odds ratio	95% confidence interval
Diabetes	1.190	(0.912,1.552)
Sex (female)	0.923	(0.807,1.056)
Age	1.080	(1.072,1.087)
Cigarettes	0.999	(0.994,1.003)
Cholesterol	1.003	(1.002,1.003)
Period 2	1.500	(1.399,1.599)
Period 3	1.654	(1.497,1.828)
Diabetes × sex (female)	1.395	(0.960,2.029)
Sex (female) × cigarettes	0.993	(0.985,1.001)

significant difference in the probability of hypertension across data collection periods. The positive signs of the period coefficients suggest the probability of hypertension is expected to be higher for subjects observed at period 2 and at period 3, compared to the baseline period of observation.

We can interpret effects very generally in terms of expected increases or decreases in the likelihood of hypertension. The coefficient for the age effect is positive and significant, meaning that older populations are expected to have a significantly higher probability of showing evidence of hypertension than younger populations. This interpretation is made accounting for the autocorrelation in the repeated observation of the same individuals, and also accounting for the effects of all of the other predictors in the model, meaning the difference in age is determined to be significant in addition to all differences described by these other effects.

Similarly, populations with higher cholesterol are expected to have a significantly higher probability of showing hypertension than similar populations with lower cholesterol. There is no evidence of a difference in the chance of showing evidence of hypertension between males and females, accounting for all of the other variables in the study. This may mean that the differences between males and females are very small, or that those differences are explained by other predictors in the model.

We can make more rigorous interpretations of predictors using the odds ratio interpretation, in terms of the odds of hypertension versus no hypertension. For age, $\exp(0.077) \approx 1.080$ indicates that the odds of hypertension is expected to increase by a multiple of 1.080 (or by $(1.080 - 1) \times 100\% = 8\%$) for an increase of one year in age, accounting for the repeated observation of individuals over time, and holding all other predictors constant. The associated 95% confidence interval suggests the true change in odds is expected to be within the range $(1.072, 1.087)$. Because we applied a marginal longitudinal model, our interpretations must be population-averaged interpretations. That means the interpretation of age is in reference to two populations that are similar with respect to all predictors but differ, on average, by one year. However, the age effect should not be interpreted as the expected change in likelihood of hypertension as an individual ages.

For the cholesterol effect, $\exp(0.003) \approx 1.003$ implies that the odds of hypertension is expected to increase by a multiple of 1.003 (or 0.3%) for an increase of 1 mg/dL of cholesterol, assuming all other predictors are constant and accounting for the repeated observation of individuals. For a more realistic change of, say, 20 mg/dL of cholesterol, the odds of hypertension is expected to increase by a multiple of $\exp(20 \times 0.003) \approx 1.062$, or 6.2%.

8.5.6 Model Prediction

Based on the longitudinal logistic regression model, the following prediction function is produced:

$$\ln\left(\frac{\hat{\pi}_{\text{Hyp}}}{1 - \hat{\pi}_{\text{Hyp}}}\right) = -5.168 + 0.174(\text{diabetes}) - 0.080(\text{sex}) + 0.077(\text{age})$$
$$- 0.001(\text{cigarettes}) + 0.003(\text{cholesterol})$$
$$+ 0.333(\text{diabetes} \times \text{sex}) - 0.007(\text{sex} \times \text{cigarettes}) \quad (8.7)$$
$$+ 0.403(\text{period 2}) + 0.503(\text{period 3}),$$

where $\hat{\pi}_{\text{Hyp}}$ indicates the estimated probability of hypertension using all of the predictors in the model. Given values of all predictors, this function can be written to estimate values of probability of hypertension:

$$\hat{\pi}_{\text{Hyp}} = \frac{\exp(-5.168 + 0.174(\text{diabetes}) + \cdots + +0.503(\text{period 3}))}{1 + \exp(-5.168 + 0.174(\text{diabetes}) + \cdots + +0.503(\text{period 3}))}. \quad (8.8)$$

Table 8.5 shows predicted probabilities of showing evidence of hypertension, for both males and females, from ages 40 through 70. Probabilities are predicted for each period of observation, and using average observed values of cigarette use and total cholesterol, assuming no presence of diabetes.

Table 8.5 *Framingham Heart Study predicted probabilities of hypertension for males and females aged 40 through 70, for each period of the study.*

		Age						
		40	45	50	55	60	65	70
Males	Period 1	0.180	0.244	0.321	0.410	0.504	0.599	0.686
	Period 2	0.248	0.326	0.415	0.509	0.603	0.691	0.766
	Period 3	0.267	0.348	439	0.534	0.627	0.712	0.784
Females	Period 1	0.161	0.219	0.292	0.377	0.470	0.565	0.656
	Period 2	0.223	0.296	0.381	0.475	0.570	0.660	0.740
	Period 3	0.241	0.317	0.405	0.500	0.594	0.682	0.759

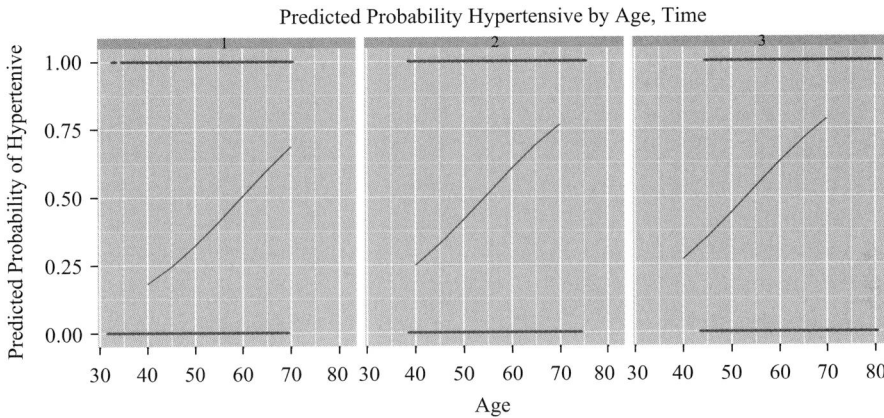

Figure 8.4 Framingham Heart Study smooth curves of predicted probability versus age, for each study period.

The predicted probabilities show that males have a higher expected chance of showing hypertension than females, accounting for the repeated observation. Predicted probabilities are clearly higher for older individuals, but also for individuals of the same age in later periods of observation. This can possibly be taken to mean that hypertension is becoming more prevalent in this population over time, regardless of age. Figure 8.4 shows a smooth estimated predicted probability curve for males across age, at each period of observation.

We applied a marginal longitudinal logistic regression model for the probability of hypertension, because we were interested in making population-averaged conclusions by comparing across different populations. The generalized estimating equations method was applied, allowing for us to determine the exact nature of the associations among outcomes for the same individual. Based on the model we found that older individuals have a higher likelihood of showing evidence of hypertension, and that individuals with higher total cholesterol have a greater chance of hypertension. The probability of hypertension showed an increase across periods of observation in the study, possibly suggesting a general increase in the likelihood of hypertension in this population over time. There was not compelling evidence of meaningful differences in the probability of hypertension between males and females, nor depending on cigarette use or presence of diabetes.

8.6 Fire-Climate Analysis: Decade Counts

We are interested in using the Fire-Climate Interactions data to predict the typical number of fire indications recorded over time. Specifically, we would like to model the decade-wise counts of fire events over the range of decades from 1130s through 2000s, while controlling for differences between regions, and also variation among sites at different regions. Therefore we will build a model using the following predictors.

- Decade: the decade, from 1130 through 2000
- Region: the region (Pacific Northwest, Interior West, and Southwest)
- Site: the site within each region

The observed sites include the Frosty, Nile Creek, and Twenty Mile sites within the Pacific Northwest region; the Ashenfelder, Cheesman Lake, Manitou, and Old Tree sites within the Interior West region; and the Blacks Mountain, Round Mountain, and the combination of the Cerro Bandera North, Hoya de Cibola, and La Marchanita sites within the Southwest region.

In this case we have a specific interest in determining how the sites change over time, and therefore a conditional longitudinal regression model is appropriate. We have selected the conditional modeling approach so that we can make subject-specific conclusions about individual sites using estimates obtained from the model.

The choice of conditional model applied involves a number of decisions. We will need to identify sources of random variation that can be explained using available variables, we will need to consider the specific components of our model that will be affected by these sources of random variation, and we will also need to consider the nature of our outcome of interest. We will begin by presenting some exploratory analysis results, followed by a number of proposed conditional models involving the identified sources of random variation. We will then evaluate the relative fit of our modeling options, make a final model selection, and proceed to interpret coefficient estimates and make predictions using the chosen model.

8.6.1 Exploratory Data Analysis

While some exploration of the Fire-Climate Interactions data was provided in Chapter 1, we provide some additional descriptive statistics here to explore the longitudinal nature of decade counts. Figure 8.5 shows the histogram and box plot associated with decade counts of fire indications. The histogram, shown in the left panel, shows a large number of decade counts near 10 (around one fire per year) and near 20 (around two fires per year). The histogram is clearly skewed to the right, as expected for count data. The box plot, shown in the right panel, provides similar information, showing that most of the values are gathered around 10, with a number of potential outlier observations extending beyond the whiskers of the box plot.

Table 8.6   *Fire-Climate Interaction average decade fire counts by site and region.*

Region	Site	Average decade fire count	
Pacific Northwest	Frosty	19.597	17.527
	Nile Creek	16.791	
	Twenty Mile	16.194	
Interior West	Ashenfelder	11.261	11.264
	Cheesman Lake	10.886	
	Manitou	11.398	
	Old Tree	11.511	
Southwest	Blacks Mountain	13.026	13.765
	Round Mountain	10.372	
	Cerro / Cibola / Marchanita	17.812	

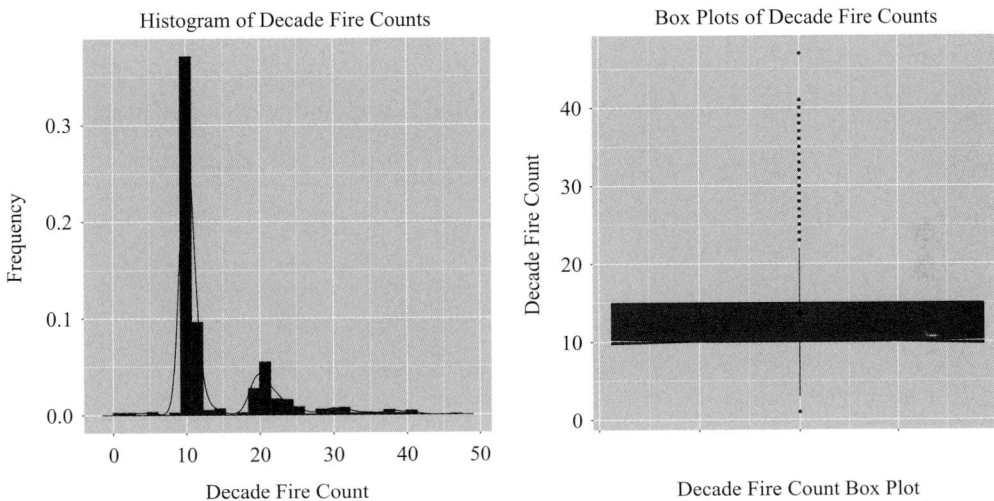

Figure 8.5 Fire-Climate Interactions decade fire counts histogram (left panel) and box plot (right panel).

Table 8.6 shows the average counts observed by site and by region. The Pacific Northwest region shows the greatest average decade counts, while the Interior West region shows the lowest. The Interior West region shows the most consistency among its site averages, while the Southwest region appears to fluctuate the most.

Figure 8.6 shows box plots of decade count for each region in the left panel and for each site in the right panel. Box plots by region support the conclusions from the table of averages. The Interior West region shows the least amount of fluctuation in decade counts, the Pacific Northwest region shows the most consistently large values, and the Southwest region shows the greatest number of unusually large counts compared to the region average. Box plots by site show discrepancies in the amount of fluctuation in decade counts, but consistent with the associated region of each site.

The left panel of Figure 8.7 shows a scatter plot of decade counts by decade, which indicates a general increase in decade fire counts through around 1750 to 1900, followed

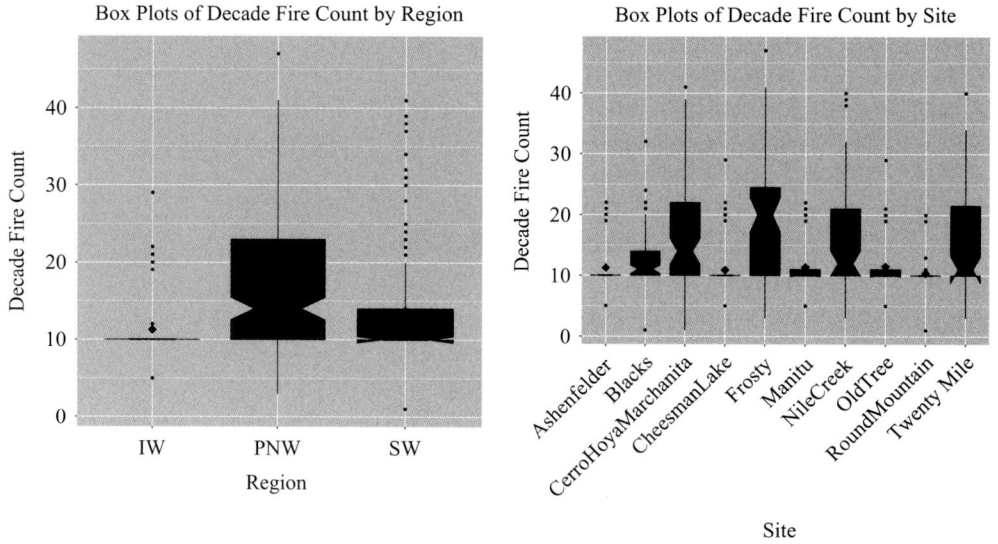

Figure 8.6 Fire-Climate Interactions decade fire counts box plots by region (left panel) and box plots by site (right panel).

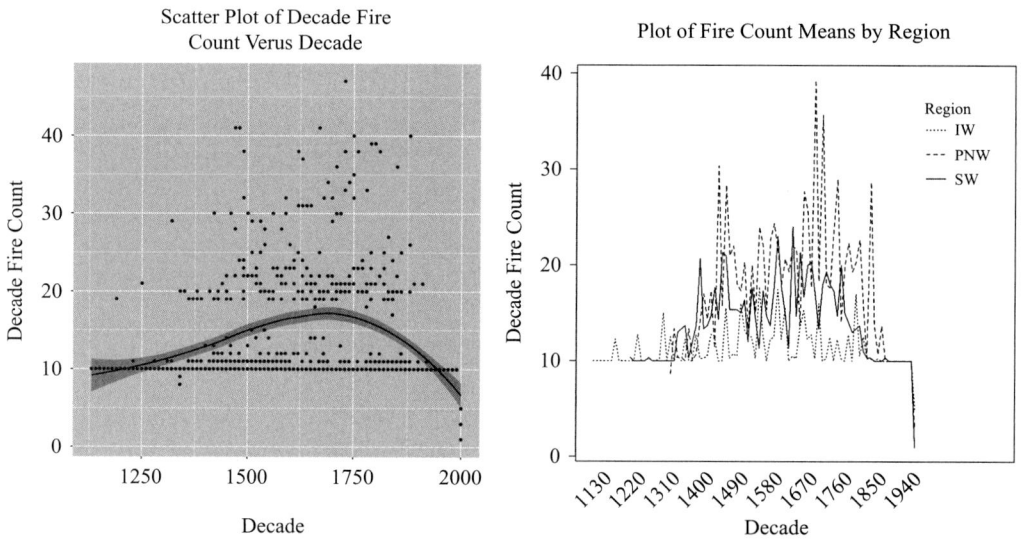

Figure 8.7 Fire-Climate Interactions scatter plot of decade fire counts by decade (left panel), line plot of average decade fire counts by region versus decade (right panel).

by a decrease in decade counts. This increase followed by decrease suggests a parabolic trajectory of decade counts. This scatter plot also shows an increase in the variation in decade counts over time. The right panel of Figure 8.7 shows the average decade fire count for each region over time. We can see that the Pacific Northwest region has the most extreme decade fire counts, while the Southwest region remains relatively consistent in the middle of the two regions.

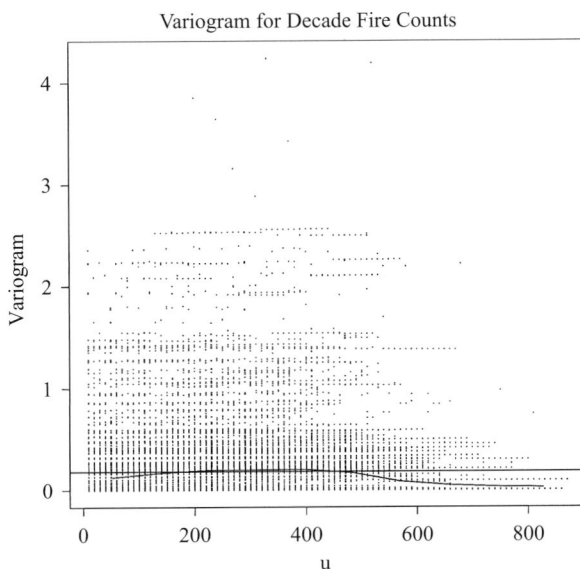

Figure 8.8 Fire-Climate Interactions variogram of residuals from decade fire count model.

8.6.2 *Autocorrelation in Decade Counts*

Before analyzing the data, it is helpful to explore the nature of the autocorrelation in decade fire counts over time. Using residuals from an uncorrelated Poisson regression model, it is possible to create a correlation matrix and explore scatter plots of residuals across time lags. However, in this case there are 88 different decades under consideration, giving 87 possible times lags to investigate. This results in a very large number of correlation values and scatter plots that would need to be investigated to determine whether a pattern is present.

Instead we use the variogram to describe the autocorrelation in decade fire counts. Figure 8.8 shows the variogram for the data, with a loess fit included to help identify any trend. Residuals were produced using an ordinary Poisson regression model, with decade and region as predictors. The estimated curve is almost completely flat until a time lag of around 500, indicating that an exchangeable or compound symmetry correlation structure may be appropriate. This is the correlation structure imposed by a random-intercept longitudinal model.

8.6.3 *Conditional Models for Decade Counts*

Our interest is in modeling the trends in decade fire counts over time for different sites, so in order to make such subject-specific conclusions we considered a conditional longitudinal approach. The most important initial step is for us to identify the sources of random variation to be included in the model. While the primary interests are in estimating trends over time, and in comparing these trends across regions, it is the *site* that represents a source of random variation. We believe the decade fire counts fluctuate randomly among the population of sites. Therefore any random effects in our model will be associated with variation among sites.

It is also important for us to consider the nature of the response. Because we are interested in modeling the decade fire *counts* as the outcome of interest, we should apply a count data model. Using count regression models, we will be able to evaluate the effect of time (decade) on the expected count, and also compare expected counts between different regions. Specifically, we investigate both Poisson and negative binomial count models, both with random intercepts and random slopes. The random intercept models can be written

$$\ln(\mu_{it}) = (\beta_0 + u_i) + \beta_1(t) + \beta_2(t)^2 + \beta_3(\text{PNW})$$
$$+ \beta_4(\text{SW}) + \beta_5(t \times \text{PNW}) + \beta_6(t \times \text{SW}), \tag{8.9}$$

where μ_{it} indicates the mean decade count for site i during decade t, u_i indicates the random fluctuation in the intercept or baseline decade fire counts among sites, PNW is an indicator of Pacific Northwest sites, SW is an indicator of Southwest sites, and a quadratic decade term has been included to capture the changing trend in decade fire counts over time. Note that we have used "t" to indicate decade for ease of subscript notation. The interaction terms between decade and region indicators allow the decade effect to differ across regions. The logarithmic transformation of the mean is applied to account for the inherent right-skewness in the decade counts being modeled. This equation can be used for any type of mean count model, including both the Poisson and the negative binomial models. Recall that using such a count model is equivalent to writing the mean decade fire counts in terms of the predictors, using the natural exponential function

$$\mu_{it} = e^{\left((\beta_0 + u_i) + \beta_1(t) + \beta_2(t)^2 + \beta_3(\text{PNW}) + \beta_4(\text{SW}) + \beta_5(t \times \text{PNW}) + \beta_6(t \times \text{SW})\right)}. \tag{8.10}$$

The exploratory investigation of the autocorrelation in the data suggested a compound symmetry model would be appropriate. Compound symmetry, in which decade counts are assumed to be correlated equally for the same site, regardless of time lag, is induced by applying the random-intercept model. However, we also considered the relative fit of a random-slopes model, in which both the linear and the quadratic components of the decade effect were allowed to vary randomly among sites,

$$\ln(\mu_{it}) = (\beta_0 + u_{0i}) + (\beta_1 + u_{1i})(t) + (\beta_2 + u_{2i})(t)^2 + \beta_3(\text{PNW}) + \beta_4(\text{SW})$$
$$+ \beta_5(t \times \text{PNW}) + \beta_6(t \times \text{SW}), \tag{8.11}$$

where the term u_{0i} represents the random fluctuation of baseline decade fire counts among sites, u_{1i} represents the random fluctuation of the decade effect among sites, and u_{2i} represents the random fluctuation among sites of the parabolic shape in fire counts across decades. This equation can be used for any count regression model, including the Poisson and the negative binomial models.

8.6.4 Conditional Longitudinal Model Fit

Model fit will be evaluated using overall model fit statistics such as AIC and model deviance, and also using the distribution of residuals. Figure 8.9 shows plots of conditional deviance residuals versus fitted values for two random-intercept Poisson count regression models. The first panel shows the deviance residual plot for the model without the quadratic decade term included, and the second panel shows the same residual plot when the quadratic

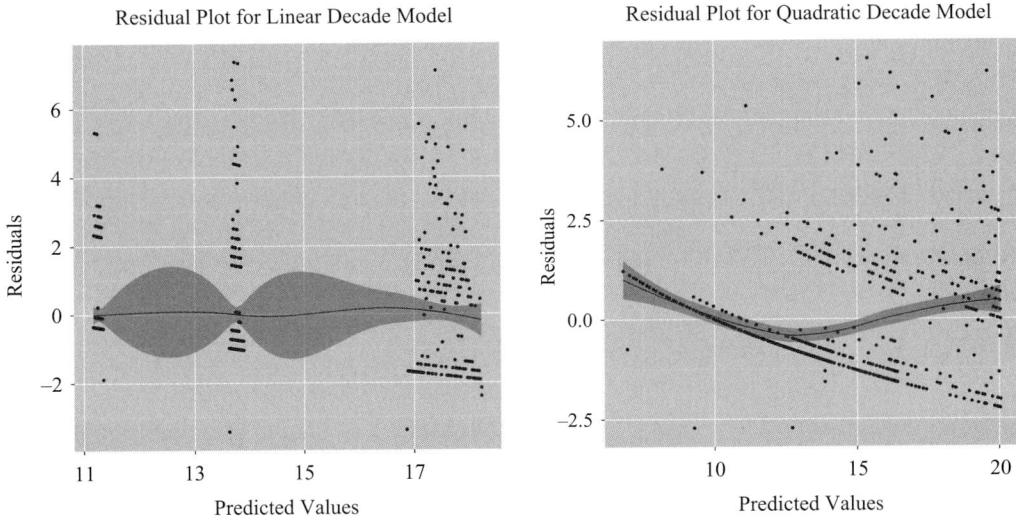

Figure 8.9 Fire-Climate Interactions scatter plot of residuals predicted values for random-intercept count model with linear decade effect (left panel), for random-intercept count model with linear and quadratic decade effects (right panel).

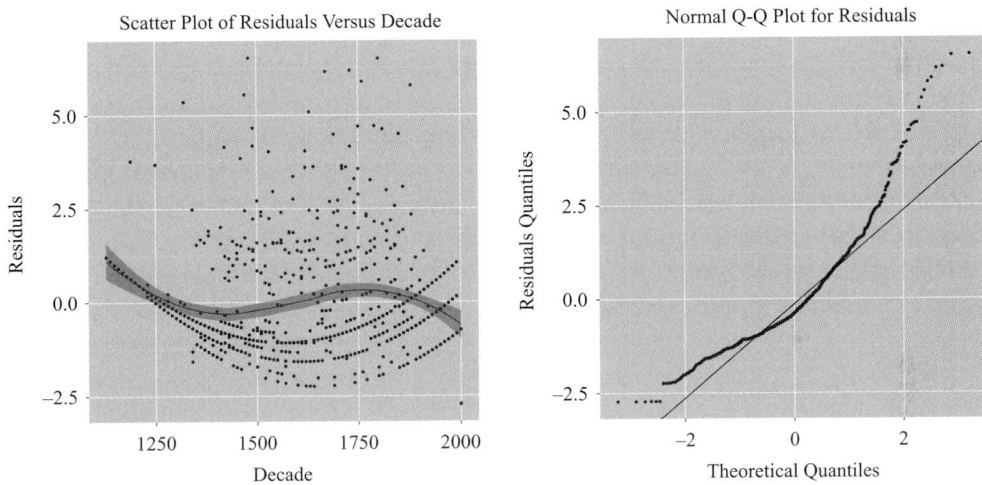

Figure 8.10 Fire-Climate Interactions random-intercept count model scatter plot of conditional deviance residuals versus predicted values (left panel), conditional deviance residuals normal Q-Q plot (right panel).

decade term is included. Clearly the quadratic time term allows the model to account for more of the dynamic nature of the relationship between decade fire counts and time, as the right panel residual scatter plot shows more consistent variation along with smaller residuals in general. The quadratic term will be included in all subsequent models.

Figure 8.10 shows two plots: the plot of conditional deviance residuals versus decade in the left panel and the normal probability plot for the same deviance residuals in the right panel, both plots using residuals from the random-intercept Poisson count regression model,

Table 8.7   *Fire-Climate Interaction model fit statistics for conditional longitudinal count model options, including both Poisson and negative binomial random-intercept and random-slopes models.*

Model	Deviance	Degrees of freedom	Deviance / DF	AIC
Poisson,				
random-intercept	1 498	770	1.945	1 514
random-slopes	1 500	765	1.961	1 526
Negative binomial,				
random-intercept	4 550	769	5.917	4 580
random-slopes	4 519	772	5.854	4 543

including the quadratic time term. The residual scatter plot shows reasonably consistent variation; however, there is evidence of patterns in the lower half of the plot. Specifically, the residual plot shows "bands" of values that are likely associated with the most common counts of 10, 20, and 30, each apparently underestimated as evidenced by the generally negative residuals on the bands. The normal probability plot shows some evidence of right-skewness, as expected for count data. The model appears to produce underestimated predictions for some of the largest observed counts.

If the random-intercept Poisson count regression model is augmented to include random fluctuation of both the linear and quadratic terms for decade, the residuals change almost imperceptibly. The new variance components for the linear and quadratic time effects in the random-slopes Poisson model are extremely small and nonsignificant. As shown in Table 8.7, the ratio of deviance to degrees of freedom increases slightly and the AIC increases from 1 514 for the random-intercept model to 1 526 for the random-slopes model. Together these values suggest the model allowing random fluctuations in the effects of decade does not produce an improved fit to the data.

Both the random-intercept and random-slopes models were re-fitted using a negative binomial distribution in place of the Poisson. While likelihood-based model fit statistics such as AIC and deviance have unclear interpretations when used to compare models with different likelihood functions, the ratio of deviance to degrees of freedom can still be compared to a value of 1 to evaluate fit. Table 8.7 shows the corresponding ratio of deviance to degrees of freedom to be farther from 1, and therefore worse, for the negative binomial models as compared to the Poisson models. We will use the random-intercept Poisson model to make conclusions and predictions.

8.6.5 Model Parameter Interpretations

Effects estimates and corresponding hypothesis tests for the random-intercept Poisson model are shown in Table 8.8. Most effects show significance using the Wald z-tests. For example, the model provides evidence that expected fire counts for sites change across decades, and the estimate of 0.012 suggests this change is an increase in fire counts. The significance of the quadratic decade term suggests there is a meaningful change in the trend of fire counts across decades, and the estimate of -0.000004 suggests the trend changes from an increase in fire counts to a decrease in fire counts, consistent with a negative parabolic shape. In other

Table 8.8 *Fire-Climate Interaction results of random-intercept conditional longitudinal Poisson regression of decade fire Counts.*

Coefficient	Estimate	Std. error	z-value	$Pr(>\|z\|)$
Intercept	−6.629	0.543	−12.216	< 0.001
Pacific Northwest	−0.673	0.205	−3.293	0.0009
Southwest	−0.361	0.180	−2.011	0.044
Decade	0.012	0.0007	17.059	< 0.001
Decade^2	−0.000004	< 0.001	−17.117	< 0.001
Pacific Northwest × Decade	0.0006	0.0001	5.162	< 0.001
Southwest × Decade	0.0003	0.0001	2.923	0.003

Random-intercept variance	0.003
Std. error	0.058

words, the model predicts that decade fire counts for an individual site follows a general pattern of increasing over time, followed by a peak and a decrease over time.

An interesting result is the negative sign of the parameter estimates for the Pacific Northwest and Southwest regions. Because the Interior West is the omitted region, both tests and parameters represent comparisons to the Interior West as a reference group. But descriptively, we found both the Pacific Northwest and Southwest regions to have *higher* average fire counts than the Interior West region, apparently contradicting the negative estimates. This is caused by the inclusion of interactions with the decade effect, both of which are positive. This implies that the Pacific Northwest and Southwest regions are expected to begin at lower decade fire counts than the Interior West region, but then increase at a greater rate than the Interior West region as decades pass. In fact this pattern is seen in the right panel of Figure 8.7, where the Pacific Northwest and Southwest regions begin with lower fire counts than the Interior West region, but increase to greater peaks in fire counts as the decade increases.

Finally, the fluctuation in baseline values among sites is estimated to be 0.003; however, the standard error suggests there is more uncertainty in this value than magnitude, so it is not clear that this variance component is significant. This does not imply that the random intercept should be removed from the model, because it is used to account for the repeated observation of the same sites over time.

8.6.6 *Model Prediction*

Predicted decade fire counts can be made according to the prediction function

$$\hat{\mu}_{it} = \exp\left(-6.629 + 0.012(t) - 0.000004(t)^2 - \cdots + 0.0003(\text{decade} \times \text{SW})\right), \quad (8.12)$$

where $\hat{\mu}_{it}$ indicates the predicted decade fire count at decade t for site i. Table 8.9 shows the predicted decade fire count means for all three regions at 50-year intervals from 1500 through 1800. The table of predicted values shows the growth and then decay in expected fire counts as the decades pass through the crest of maximum fire counts seen in the scatter plots, occurring somewhere between 1600 and 1700. Values also show predictions to be

Table 8.9 *Fire-Climate Interaction conditional longitudinal model prediction of decade fire counts from 1500 through 1800, by region.*

Region	Decade						
	1500	1550	1600	1650	1700	1750	1800
Pacific Northwest	18.356	19.262	19.834	20.043	19.875	19.341	18.469
Southwest	15.675	16.192	16.415	16.329	15.941	15.271	14.356
Interior West	13.805	14.030	13.994	13.696	13.155	12.398	11.468

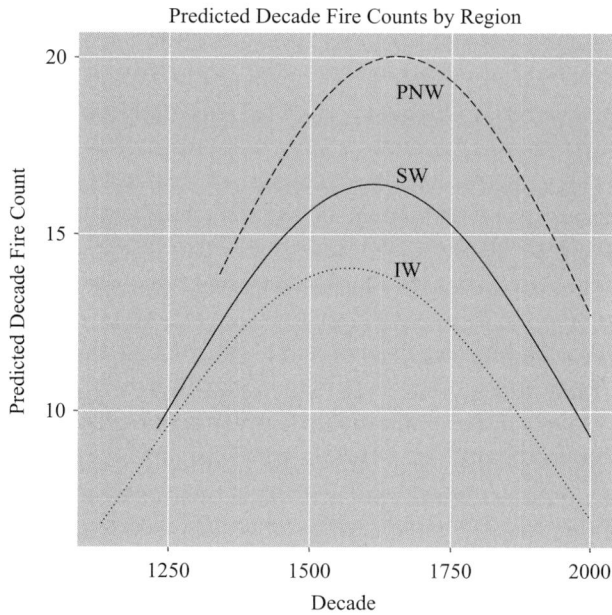

Figure 8.11 Fire-Climate Interactions random-intercept count regression model predicted decade fire counts by decade, for each region.

greatest for the Pacific Northwest region, and smallest for the Interior West region, also consistent with exploratory analyses.

Figure 8.11 shows predicted decade fire counts across all decade from the data, with separate smooth prediction lines for the three regions under consideration. The plot clearly shows the peak of predicted fire counts around the years 1600 to 1700 for each region, followed by a dramatic decrease. Predictions also show that the Pacific Northwest region is expected to show greater numbers of fire counts than the other two regions across all decades, while the Interior West region is expected to show lesser counts.

We applied a conditional longitudinal count regression model to allow us to make conclusions about the trends in decade fire counts over time, allowing for the site to represent a source of random variation. We found the random-intercept Poisson model, in which the baseline value of mean fire counts fluctuates randomly among the sites, to be the most appropriate for these data. Such a model imposes an exchangeable working correlation structure on the data, so that decade fire counts for the same site are assumed to be correlated equally, regardless of the time lag between observations. The analysis showed decade fire counts to be effectively modeled using a quadratic relationship, with significant differences in this relationship between regions of observation. Regions appeared to show their greatest decade fire counts between 1600 and 1700, followed by noticeable decreases. The Pacific Northwest region showed the largest fire counts, while the Interior West region showed the lowest.

8.7 Summary

In this chapter we investigated models for data collected from the same individuals or objects over time, referred to as longitudinal data. For such data we must apply models that allow for the inherent nonindependence of responses, which is assumed in the normal linear regression model. We discussed and applied both marginal longitudinal models and conditional longitudinal models. Marginal longitudinal models are appropriate when the analyst has an interest only in making population-averaged conclusions, in which expected outcomes are compared across populations. Such estimation approaches, including the generalized estimating equations, allow the data analyst to directly specify the nature of the associations among responses over time. Conditional longitudinal models can be used to make population-averaged conclusions, but more easily allow for subject-specific conclusions, in which individual trends or changes over time are of interest.

We modeled the probability of hypertension over time using a marginal longitudinal logistic regression model. This model allowed us to make comparisons of the probability and odds of hypertension between populations, while properly accounting for the inherent associations among repeated observations of hypertension. Applying the generalized estimating equations approach allowed us to evaluate a number of possible correlation structures, after which we identified the autoregressive structure to be most appropriate. From the analysis we found older populations and those with higher total cholesterol to have a higher likelihood of being hypertensive. We also found the probability of hypertension to increase across periods of the study.

We modeled the expected decade fire counts over the decades 1130 through 2000 using a conditional longitudinal Poisson count regression model. We chose this model because it allowed us to make statements about trends of decade fire counts over time for the same sites, which were identified as a source of random fluctuation. Using the random-intercept conditional model, we allowed the baseline values of decade fire counts to vary randomly among sites and implicitly imposed an assumption of equal correlation in decade fire counts within sites. Based on the analysis, we found a quadratic trend in decade fire counts, in which the fire counts rose to a peak and then decreased, to be appropriate and to differ across regions.

8.8 Further Reading

We have found Diggle et al. (2002) to be the most comprehensive resource on longitudinal data analysis techniques. We have also found Fitzmaurice et al. (2012) to be extremely thorough. The text with the most accessible model interpretations may be Hedeker and Gibbons (2006). A nice, brief overview of longitudinal data methods and interpretations is given in Zeger and Liang (1992). An excellent starting point would be Fitzmaurice et al. (2008), in particular their chapter on the history of longitudinal data analysis. A very understandable and application-focused discussion is given in Twisk (2013).

An excellent and easy-to-understand treatment of the generalized estimating equations is given in Hardin and Hilbe (2003). A more abstract discussion is given in Ziegler (2011).

Twisk (2006) provides very understandable explanations of coefficients in conditional longitudinal models. A very high-level treatment of correlated models is presented in Lee et al. (2006).

9

Structural Equation Modeling

9.1 Introduction

Structural equation modeling (SEM) may be considered as using models of covarying variables. The models systematize variables by type, either manifest (observable and measurable variables) or latent (not directly measurable variables), that depend on the covariance of other manifest or latent variables. This includes the popular regression models of manifest variables dependent on covarying manifest predictor variables. An extension of regression models allows for the analysis of latent variable effects underlying manifest predictor variables that influence a manifest response variable. Latent prediction variables characterizing a response leads to models in which one or more latent variables that influence the covariance behavior of a set of manifest variables. Finally, SEM includes models of latent variables influencing not only manifest variables, but also other latent variables. SEM, then, is a set of models designed to manage a variety of relationships among both observable and unobservable variables.

Prior to utilizing SEM, analysts and researchers must formulate a hypothesis relating the variables of interest. That is, SEM requires subject matter expertise to devise meaningful variable covariance structures. The outcome of SEM analysis is assessments of chosen data set substantiation of the proposed hypotheses. The validity of SEM lies in the unification the analysis and the subject matter.

Intelligence quotient measures, motivation level identification, and ancient crater-producing impactor properties are examples of problems that can be examined with SEM. Each of these examples of latent behavior and resurfacing agents require the quantification of proxy or aftermath observable measurements, and are investigations of proposed or hypothesized relationships among a set of concomitant variables. Intelligence quotients often cannot be measured directly, so other related variables such as quantitative reasoning, reading and writing ability, short- and long-term memory, and visual processing are measured and hypothesized to be the outcomes of levels of intelligence. Purchasing motivation may be hypothesized to be the influencer of internet link paths among websites including the website link sequence and the website dwell time. Scientists hypothesize and have sample evidence that asteroids and comets caused impact craters on the Moon. Variables that can be measured to associate the observed impact crater characteristics with the causal impactor latent unobserved characteristics are the observed orbiting asteroids and comets, existing crater diameters, rim-to-floor depths, and ejecta spatial patterns.

In each of these three examples, manifest variables are due to the influence of latent variables. When a manifest variable's measures are the result of the influence from one or more latent variables, the manifest variables are called indicator variables of the latent variable characteristics (not to be confused with regression indicator variables which are also known as dummy variables). Having methods to evaluate variables that are not measurable provides investigators additional analytic capability. Problems often treated with qualitative analyses thus may be supplemented with quantitative analysis using SEM.

SEM is comprised of a structural component, a variable relationship path diagram component, and a latent variable component when latent variables are of interest. A structural component in SEM is regression models. Variable path diagrams show the direction of influence, which, for a simple regression structure, divides the manifest variables into a response variable and a set of predictor variables, with path arrows pointing from the predictors to the response. The latent component has been described above with the intelligence quotient, motivation, and impact crater examples. When the structural component is not appropriate for a particular model, but the latent variable component is, the model is known as a latent variable model. The three examples above are latent variable models (LVMs). The correlation structure hypothesized among the variables is essential to LVMs. We employ both the LVM and the structural SEM using the Framingham Heart Study data, and the School Study of Crime and Safety data, respectively.

SEM analysis requires the specification of a model with known variable relationships. The model effects and model fit are estimated to ascertain the significance relating the variables. As the SEM specifies knowledge of the relationship among the variables, confirmatory factor analysis, known as CFA, may be used. The form of LVMs in the example analyses below as well as the three examples given above is termed a reflective latent variable model. The reflective latent variable model implies the latent variables *influence* the indicator variables' covariance behavior. The form of SEM latent variable models not in this chapter is that of formative latent variable models in which the latent variable variances are the *result* of the covariation the indicator variables.

SEM considers that covariant relationships are free from measurement error leaving only the variance common to all the variables. Measurement error is accounted for by specifying or estimating the size of the error from, for example, gauge reputability and reliability studies, or internal consistency studies.

The purpose of this chapter is to provide a simple introduction with applications of SEM, so we strongly recommend a thorough review of the references listed at the end of this chapter to become proficient with the multitudinous capabilities of SEM.

9.1.1 SEM Variable Categories

SEM methods categorize variables into variables that behave without causal activity, and variables that behave as the outcome of one or more influencing actions. Within any particular study, variables with behaviors resulting from noncausal action are known as exogenous variables. Variables that behave due to causal actions are called endogenous variables. In regression, for instance, endogenous variables are the response variables, e.g., the counts of fire-scarred tree rings respond to the location from which the ring sample was taken. The predictor variables of region and decade are the exogenous variables. For regression, then, endogenous variable behavior is influenced by exogenous variables whose behavior is not driven or caused by variables within the study. In the reflective

LVM examples below, the exogenous variables are the latent variables, and the endogenous variables are the indicator variables.

9.1.2 Model Types

There are three model types in SEM discussed here: LVM, full SEM, and seemingly unrelated regression (SUR), though this chapter is restricted to LVM and SEM. SUR is discussed briefly below due to its common use in stock market analyses. LVM is a model type in which a set of latent variables influence the covariance of a set of indicator variables. Full SEM is a LVM with a regression equation component. For example, as seen below with the School Study on Crime and Safety data, a LVM in which the latent variable called "school climate" influences indicator variables including the amount of crime and the number of suspensions. Also assumed is a LVM in which the latent variable "academic achievement" influences indicator variables such as English language competency and test scores. The hypothesis is that the LVM of academic achievement regresses on the LVM of school climate. The two LVM structures along with the regression between these two LVM structures constitute a full SEM.

The seemingly unrelated regression (SUR) models may be considered a special case of the regression models as addressed in Chapters 2 and 3. SUR is assumed to not share independence of residuals among models as with normal regression models. The estimators derived from ordinary least squares are still valid, however, if the regression residuals are shown to be normal. These are not guaranteed, however, so EDA and model diagnostics are essential.

The basic SUR model assumes a single regression relationship with a set of predictors for a set of responses, or, for multiple sets of responses each with an associated (often different) regression relation. This response set of regressions may be solved simultaneously producing parameter estimates for each regression relationship. Thus, a set of simultaneous linear equations in which the predictors per response can be unique or can be shared among responses is analyzed with ordinary least squares estimation. Recall from Chapter 2 the example of two companies with data over the same period for capital stock prices, outstanding shares values, and gross investments. The two companies depend on the products of the other for their respective product manufacturing processes. The gross investment of each company as a function of their respective outstanding shares values and capital stock prices is modeled as two regression equations analyzed as a linear system of equations. The analysis includes an examination of the two models' residuals correlations to measure the amount of endogenous association. Significant residuals correlations suggest the gross investment for one company is highly dependent upon the gross investment of the other company, and vice versa.

9.1.3 SEM Paths

Assuming a hypothesis about the relationship among endogenous and exogenous variables, a path diagram depicts the type of relationship between any pairing of these variables. Note that variability within a variable is depicted as self-influencing. The symbols used to represent the variables and their relationships are given in Table 9.1.

Table 9.1 *SEM path analysis symbols.*

SEM path analysis symbols	
Symbol	Description
Rectangle	Manifest variable
Circle	Latent variable
Triangle	Constant or (1)
Single-ended arrow	Direct influence. Usually a straight arrow.
Double-ended arrow	Covariance. Often a curved arrow.

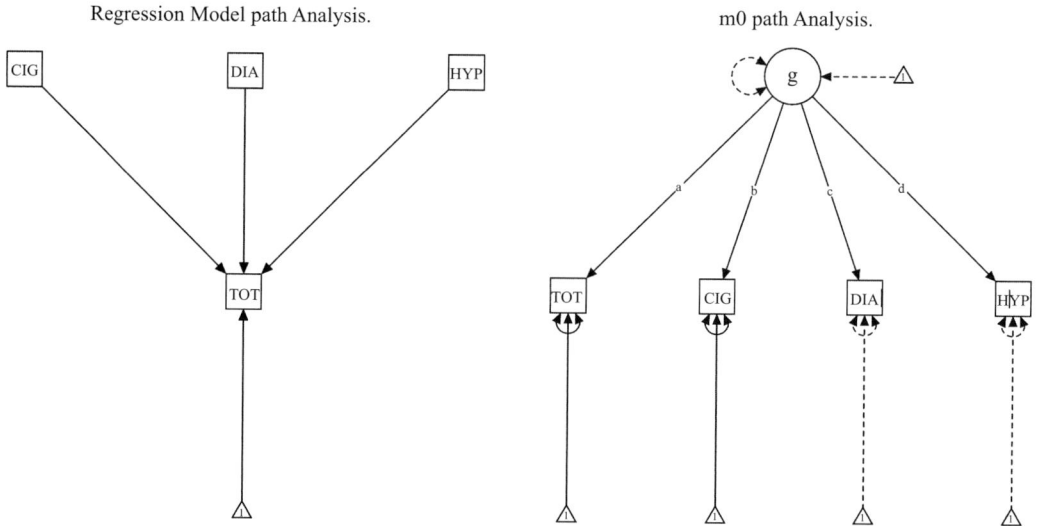

Figure 9.1 The left-hand panel is the path analysis for a multiple regression of total cholesterol (TOT) on number of cigarettes used per day (CIG), presence of diabetes (DIA), and presence of hypertension (HYP). All the arrows for this regression have their points at TOT. The right-hand panel is the path for the LVM of stress (g) on total cholesterol (TOT), number of cigarettes used per day (CIG), and the presence of hypertension (HYP) and diabetes (DIA).

The left-hand panel of Figure 9.1 gives a path diagram for a multiple regression of total cholesterol (endogenous response variable, "TOT"), versus the presence of hypertension and diabetes, and the number of cigarettes used per day (exogenous predictor variables, "HYP," "CIG," and "DIA") from the FHS data. Each variable is measurable, hence the use of rectangles to represent the their roles as manifest variables. The triangle represents the model intercept which is constant for these data. All the arrows are straight, direct influencers on total cholesterol, and thus all have arrow tails originating at the predictor with arrowheads terminating at total cholesterol. A common addition to the regression-type path analysis is a circle representing the inherent error on the response, total cholesterol. The error circle is connected by a straight line to the rectangle response with the arrowhead to the rectangle. A double-ended curved arrow on the error circle depicts the within-response error.

The right-hand panel of Figure 9.1 gives a path diagram for exogenous latent variable stress (g), and the four endogenous manifest indicator variables of total cholesterol (TOT), and the number of cigarettes smoked per day (CIG), presence of diabetes (DIA), and hypertension (HYP). The single-ended, straight-line arrows are directed from the latent variable (circle) to each of the manifest variables (rectangles). The hypothesis being modeled is that stress influences the levels of total cholesterol, the number of cigarettes smoked in a day, and the presence of hypertension and diabetes. Each variable, either latent or indicator, has a curved, double-ended arrow that represents each variable's inherent error. Both the latent and indicator variables have constant intercepts represented by the respective triangles.

9.1.4 Confirmatory Factor Analysis

Factor analysis manifests as either exploratory factor analysis or confirmatory factor analysis (CFA). Exploratory factor analysis is used to discover relationships among variables, either latent or manifest or both, while confirmatory factor analysis is constrained by the assumed relationships among variables. Exploratory factor analysis requires no a priori relationship assumptions among the variables, whereas confirmatory factor analysis does. Note that the path analyses for both exploratory factor analysis and for confirmatory factor analysis are the same. What differs is the intent of the analysis. The latent variable model of stress on total cholesterol, cigarette use, and presence of diabetes and hypertension in the right-hand panel of Figure 9.1 could be either an exploratory or a confirmatory factor analysis.

The object of factor analysis is often to relate latent variables to manifest variables. As such, there is a correlation structure that is assumed to relate the manifest variables to the latent variables. The CFA assumes the covariances among the manifest variables are the result of the influence of the latent variables. Given unmeasured values of a set of latent variables, called factors, that influence the manifest variables, each of which is assumed independent from any of the other manifest variables, the correlations among the manifest variables are assumed the result of their relationships with the factors. The correlations are from linear combinations of the manifest variables to characterize the factors. The coefficients of the linear combinations are known as factor loadings.

The analyst specifies the scales on the factor variances as well as the measures of central tendency of the factors as the factors are not measured directly. For example, the data may be standardized to account for large-scale differences in the manifest variables. Standardization of the data removes the measurement scale dependence, and replaces them with scores that adjust the relative scale proportions. A common assumption on the factors is that they are independent of each other, though analytical methods can account for varying degrees of possible dependencies.

Manifest variable variance is the sum of the squares of the factor coefficients (square of the factor loadings) and the residual error from modeling the manifest variable (mean-centered) by the one or more factor loadings. The summed squares are the variances common to all the manifest variables (the communality), leaving the residual error being the unique variance of each manifest variable. The product between one manifest variable's factor loadings with the corresponding factor loadings of another manifest variable gives the covariance between these two variables. The advantage of factor analysis is that manifest variable covariation is not dependent on the manifest variables, but rather the covariation describes the influence on the manifest variables by the factors (the latent variables) as derived from the communality variance.

The factor analysis model is

$$
\begin{aligned}
x_1 &= \lambda_{11} f_1 + \lambda_{12} f_2 + \cdot \cdot + \lambda_{1p} f_p + a_1, \\
x_2 &= \lambda_{21} f_1 + \lambda_{22} f_2 + \cdot \cdot + \lambda_{2p} f_p + a_2, \\
&\ \vdots \\
x_q &= \lambda_{q1} f_1 + \lambda_{q2} f_2 + \cdot \cdot + \lambda_{qk} f_p + a_q,
\end{aligned}
\tag{9.1}
$$

where the x_i are the manifest variables and the f_j are the factors (latent variables). In general the manifest variables x_i are observed data, while the coefficients λ_{ij} must be estimated through the model. In the case of the stress model described above, there is a single factor, stress, and four manifest variables: $x_1 =$ total cholesterol, $x_2 =$ number of cigarettes used per day, $x_3 =$ presence of diabetes, and $x_4 =$ presence of hypertension. The analysis calculates the values of the coefficients λ_{11}, λ_{21}, λ_{31}, and λ_{41}; the a_i are zero-mean, finite variance errors that are assumed to be independently distributed.

To derive the effect of the latent variables on the manifest variables' covariance, the variance of each of the manifest variables, x_i, is written as

$$
\mathrm{Var}(x_i) = \sum_{j=1}^{p} \lambda_{ij}^2 + \psi_i.
\tag{9.2}
$$

The sum of the squares of the λ_{ij} is the communality of the ith manifest variable and is the variance shared among the manifest variables due to the influence of the latent variables. The ψ_i is the amount of variance in the ith manifest variable not shared with any of the other manifest variables and is called the unique variance. The manifest variables' covariances are written as

$$
\mathrm{Covar}(x_{il}, x_{jl}) = \sum_{l=i}^{p} \lambda_{il} \lambda_{jl}.
\tag{9.3}
$$

These covariances, when combined with the manifest variable's variances ψ_i, allow the estimation of these variances from the data such that the effects of the latent variables on the manifest variables is assessed.

9.1.5 Evaluating Model Fit

Several statistical tests are used to evaluate how well the data fit the hypothesized SEM. A good fit between the model and the data *does not mean the theory or hypothesis is validated*, or that the fit explains the partitioning of the covariance. Rather, a good fit suggests the hypothesis is plausible, hence the reason for deep integration of analysis and subject matter. The statistics most often used to assess model fit are the χ^2 test, the root mean square error of approximation (RMSEA), the comparative fit index (CFI), and the standardized root mean square residual (SRMR). We use these fit statistics to assess model adequacy in the modeling examples below. Several other fit statistics are available to assess the aptness of a model, and several should be considered for model fit. See Section 9.5 for references.

χ^2 *Test*

The χ^2 test is used to measure the difference between the observed covariance and the expected covariance. χ^2 values near zero suggest a smaller difference between the observed and the expected covariance which implies a good fit. The χ^2 test may fail to reject an inadequate model fit to a small sample size, and it may reject an adequate model fit to a large sample size. Thus one reason to utilize several fit statistics.

Root Mean Square Error of Approximation

The root mean square error of approximation (RMSEA) is a measure of whether a model reasonably reproduces the data used to construct the model. This may be considered the analog of the regression standard error of the estimate. The RMSEA is not affected by sample size as is the case with the χ^2 test. The RMSEA statistic ranges between 0 and 1. Smaller values are desirable with a value of 0.06 or less generally used for identifying adequate model fit.

Root Mean Square Residual and the Standardized Root Mean Square Residual

The root mean square residual (RMR), which may be transformed to become the standardized root mean square residual (SRMR), is the square root of the differences between the sample covariances and the model covariances. The values of the RMR are dependent on the scales of the indicator variables, and hence the statistic may be biased. The SRMR standardizes the indicator variables thus scaling the SRMR range from 0 to 1. A SRMR value of 0.08 or less suggests an adequate model fit.

Comparative Fit Index

The comparative fit index (CFI) measures the difference between the data and the hypothesized model after adjusting for sample size in the χ^2 test. The CFI ranges from 0 to 1, with values of 0.90 and greater suggesting adequate model fit.

It is important to reiterate that the model with adequate fit to a data set does not suggest a theory or hypothesis is correct, only the data suggest the hypothesis under consideration is plausible.

9.2 FHS Analysis: Latent Stress

We wish to examine the Framingham Heart Study (FHS) data to test if stress affects patient total cholesterol levels, the number of cigarettes smoked per day, diabetes diagnosis, and diagnosis of hypertension. The variable "stress" is treated as a latent variable which means it cannot be measured directly. The total cholesterol, the number of cigarettes used per day, diabetes diagnosis, and diagnosis of hypertension variables that are assumed to be directly influenced by the stress latent variable are the indicator variables. Thus, the hypothesis is that the latent variable produces covariances among the indicator variables. The simplistic hypothesis is, does a single factor model with latent variable structure fit the data?

Table 9.2 *Framingham Heart Study descriptive statistics for continuous variables.*

Variable	Minimum	Median	Mean	Maximum	Variance
Time of hypertension	0	2 429	3 599	8 766	11 996 207.990
Total cholesterol	107.0	238.0	241.2	696.0	2 059.309
Age	32.00	54.00	54.79	81.00	91.325
Cigarettes	0.00	0.00	8.25	90.00	148.000

Table 9.3 *Framingham Heart Study descriptive statistics for categorical variables.*

Variable	Levels	Number	Percent
Hypertension	Yes	8 642	74.3%
	No	2 985	25.7%
Sex	Female	6 605	56.8%
	Male	5 022	43.2%
Diabetes	Yes	530	4.6%
	No	11 097	95.4%

Although there are several variables in the data set, those of interest are the same as in the logistic model of Chapter 5:

• Cigarettes: number of cigarettes smoked per day, $\sqrt{\text{cigarettes}}$ used in model
• Cholesterol: serum total cholesterol (mg/dL), ln(total cholesterol) used in model
• Prevalent hypertension: prevalent hypertensive patients are excluded from the analysis
• Hypertension: diagnosed in the study period as hypertensive
• Diabetes: diagnosed with diabetes either prior to study entry, or during the study

The FHS data first had all cases in which a subject had hypertension diagnosed prior to the study start date removed as well as any case for which values of the variables listed above were missing. Table 9.2 is a summary of the continuous and count variables included in the SEM analysis. Recall that we use the natural logarithm of the total cholesterol and the square root of the number of cigarettes used per day.

Table 9.3 gives the numbers and proportions of the sex levels (1 = male, 2 = female), subjects' level of diabetes (0 = no diabetes, 1 = diabetes present), and subjects' hypertensive state (0 = no hypertension, 1 = hypertensive).

The right-hand panel of Figure 9.1 gives path analysis of the latent variable, stress (g), which is assumed, by arrowhead direction, to influence the levels of total cholesterol (TOT), the number of cigarettes used per day (CIG), onset of diabetes (DIA), and onset of hypertension (HYP). Notice the double-headed arrows on the variables which indicate the inherent errors of each variable.

Latent variable variance is estimated by assuming this variable is standardized, and thus the standardized indicator variables are used. The indicator communality variances and

Table 9.4 *Framingham Heart Study Stress LVM model measures of fit.*

FHS SEM Fit Measures			
Statistic	Estimate	Degrees of freedom	*p*-value
npar	10		
fmin	3×10^{-4}		
chisq, scaled	1.7507	2	0.4167
chisq, scaling factor	1.0235		
cfi, scaled	1		
rmsea, scaled	0		0.9968
srmr	0.0081		

unique variances then are measured in changes in standard deviations. The "stress" model is

$$TOT = \lambda_1(stress) + a_1,$$
$$CIG = \lambda_2(stress) + a_2,$$
$$DIA = \lambda_3(stress) + a_3,$$ \hfill (9.4)
$$HYP = \lambda_4(stress) + a_4.$$

The statistics we use to assess model fit is the χ^2 test, the root mean square error of approximation (RMSEA) , the comparative fit index (CFI), and the standardized root mean square residual (SRMR). The number of parameters to estimate is 10 as given in Table 9.4 as "npar." For large sample sizes such as we have with these FHS data, the sample size is multiplied by the value "fmin" which is distributed as a χ^2 statistic with degrees of freedom equal to the amount of nonredundant information less the number of estimated parameters (npar = 10). The estimated χ^2 value is 1.9562 on 2 degrees of freedom and a *p*-value of 0.3760, and we thereby fail to reject the null hypothesis of an adequate fit. This χ^2 value suggests the data are a plausible fit to the model.

The CFI has a value of approximately 1. Values close to unity suggest the data are a plausible fit to the model. The RMSEA of approximately 0 also indicates the model reasonably approximates the data as opposed to close to 1 indicating the model is a poor fit. The RMSEA *p*-value $0.9959 \gg 0.05$ is interpreted as the RMSEA is at least ≤ 0.05, which is considered as close enough to 0 to conclude the model is a close fit. As with RMSEA, SRMR $= 0.0081$ close to 0 suggests the model is an adequate representation of the data.

Each of the model fit measures assess the model fit as an adequate representation of the data. A final check is a comparison of the implied correlations from the model with the correlations of the raw data. Table 9.5 shows the lower triangular matrix of raw data correlations. If our SEM is well posed, the model-implied correlations will reflect the same correlative relationships as the raw correlations. An examination of the model-implied correlations in Table 9.6 give the same general correlation statistics as the raw data correlation in that the signs of the respective variable pairings are the same, and

Table 9.5 *Framingham Heart Study stress LVM raw data correlation matrix.*

FHS raw data correlation matrix			
	Cholesterol	Cigarettes	Diabetes
Cholesterol			
Cigarettes	−0.0119		
Diabetes	0.0252	−0.0091	
Hypertension	0.0821	−0.0847	0.0044

Table 9.6 *Framingham Heart Study stress LVM model-implied correlation matrix.*

FHS model-implied correlation matrix			
	Cholesterol	Cigarettes	Diabetes
Cholesterol			
Cigarettes	−0.0154		
Diabetes	0.0054	−0.0055	
Hypertension	0.1035	−0.1065	0.0373

Table 9.7 *Framingham Heart Study stress LVM residuals correlation matrix.*

FHS model residuals correlation matrix			
	Cholesterol	Cigarettes	Diabetes
Cholesterol			
Cigarettes	0.003		
Diabetes	0.071	−0.0219	
Hypertension	≪ 0.001	−0.001	−0.020

the magnitudes equivalent levels of correlation. Note that in each correlation matrix, only the endogenous variables are associated.

As with all the models we have presented in this handbook, the residuals from the SEM are important to assessing the model fit to the data. The residuals correlation matrix presents the among-variables correlations without the self-correlations, which do not provide model fit information. No correlation is desired for pairwise correlations, viz., we wish all the pairwise correlations to be zero. We see from Table 9.7, which is the model residual correlation matrix, that we have the desirable property that no two endogenous variable residuals are correlated; i.e., all two-way correlations round to zero. These correlation analyses suggest that the SEM is an adequate fit to the FHS data.

We now look to the response measures of the indicator variables as influenced by the latent variable. Table 9.8 lists the parameter estimates of the SEM CFA, specifically for the influence of stress on the indicator variables after using the standardization method. The

Table 9.8   *Framingham Heart Study Stress LVM parameter estimates from m0 path analysis model.*

		FHS LVM parameter estimates			
Path start	End	Estimate	Std. error	*z*-value	*p*-value
Stress	Cholesterol	0.02	0.01	1.62	0.11
	Cigarettes	−0.28	0.17	−1.62	0.11
	Diabetes	0.04	0.08	0.54	0.59
	Hypertension	0.85	0.51	1.67	0.09

Table 9.9   *Framingham Heart Study Stress LVM loadings, communality, and uniqueness.*

	FHS LVM variance allocation		
Variable	Loading	Communality	Uniqueness
Cholesterol	0.12	0.01	0.99
Cigarettes	−0.13	0.02	0.98
Diabetes	0.04	0.00	1.00
Hypertension	0.85	0.72	0.28

standardization method was used due to the presence of the dichotomous indicator variables of onset of hypertension and the onset of diabetes. The standardization method assumes the latent variable is standardized, and thus the SEM analysis standardizes the indicator variables except for the dichotomous indicator variables. We see from the *p*-value column in the upper portion of the table that all but the onset of hypertension is marginally influenced by stress at a 0.1 level. We thereby question the usefulness of this SEM hypothesis as the p-values are larger than commonly used levels. While the *p*-values and the associated *z*-scores call into question the usefulness of this SEM, the loadings and communality statistics must first be examined.

For each indicator variable, we calculate the communality which is the portion of the indicator variance attributed to the latent variable. Communality is the squared value of the standardized parameter (path) estimates of a specific indicator variable. The difference between 1 and the communality value gives the unique variability, which is the variability unique to the specified indicator variable. Table 9.9 gives the loadings for each indicator variable, and the respective communality and uniqueness estimates. The values of the unique variance suggest that the majority of the indicator variance is unique to each indicator with the exception of presence of hypertension. The values of the communality values are quite small except for the onset of hypertension. This suggests the hypothesis that stress influences total cholesterol, the number of cigarettes smoked per day, and the onset of diabetes, is likely not plausible. Interestingly, we see the communality value of 0.72 suggests stress may influence the onset of hypertension, a finding suggested by the *p*-value discussed above. Certainly, further consideration should be given to the hypothesis formulation if the FHS data are to be used to assess the influence of stress on selected indicator variables.

The LVM fit statistics suggest the stress latent variable model is a good fit to the FHS total cholesterol, number of cigarettes used per day, onset of diabetes, and onset of hypertension data. However, the variance components assessments suggest that only the onset of hypertension is influenced by stress levels. We suggest the conclusion is that the hypothesis that stress indeed may be measured by the measured values of the four indicator variables is tenuous, and further research is recommended.

9.3 SSOCS Analysis: School Climate and Academic Success

The first question to investigate using the School Survey on Crime and Safety (SSOCS) is whether school climate influences the covariance of measures of school security and undesirable behaviors, and whether academic achievement influences the covariance of measures of academic success. The second question is whether school climate and academic achievement correlate with each other. We therefore have not just one latent variable influencing several indicator variables as with the Framingham data, rather, we have two latent variable models. Additionally, we have that one of these latent variables is a regressing variable for the other latent variable. Of all the variables in the data set, we use the same variables as in Chapter 5 for indicator variables, viz.:

- Suspensions: the number of suspensions
- Uniforms: use of uniforms
- Metal detectors: use of metal detectors
- Tipline: availability of tip lines
- Counseling: availability of counseling
- Crime: area crime as low, medium, or high
- English: English language proficiency
- Below 15th: percent of students below the 15th percentile on tests

The latent variables influencing the indicator variables are school climate and academic achievement. Figure 9.2 gives the path analysis of the two latent variable models for school climate (C) which is assumed to influence the use of metal detectors (mt), use of tip lines (tp), levels of crime (crm), and number of suspensions (ssr), as indicated by the direction of the arrowheads.

Similarly, the latent variable academic achievement (A) is assumed to influence (note the direction of the arrowheads) use of uniforms (unf), use of counseling (cns), percentage of English language competency (Eng), and the percent of students below the 15th percentile on tests (b.1). Regardless of whether a variable is a latent variable or an indicator variable, the double-headed arrows looping on the variables indicate the inherent errors. The structural component is the assumed regression of academic achievement on school climate, depicted by the straight arrow directed from the school climate, "C" to the academic achievement "A."

The path analysis of the SSOCS SEM is constructed based in part on the continuous data summary from Chapter 1 (reproduced in Table 9.10 for convenience) and on data that are included in the variable list above. The data summary shows that distribution analysis is necessary as the means and medians of these continuous variables appear to be different, suggesting skew to the right, with the means larger than the medians.

Figure 9.3 has the normal Q-Q plots for the suspensions, adjusted suspensions, log-transformed limited English percentage, and the percent of students below the 15th

Table 9.10 *School Survey on Crime and Safety descriptive statistics for continuous variables.*

Variable	Minimum	Median	Mean	Maximum	Variance
Number of suspensions	0	0	7.852	3 000	4 863.123
Number of insubordinate students	0	16	88.76	9 608	118 589.792
Percent age with limited English	0.00	2.00	8.727	100	217.387
Percent age below 15th percentile on tests	0.00	10.00	13.77	100	208.417

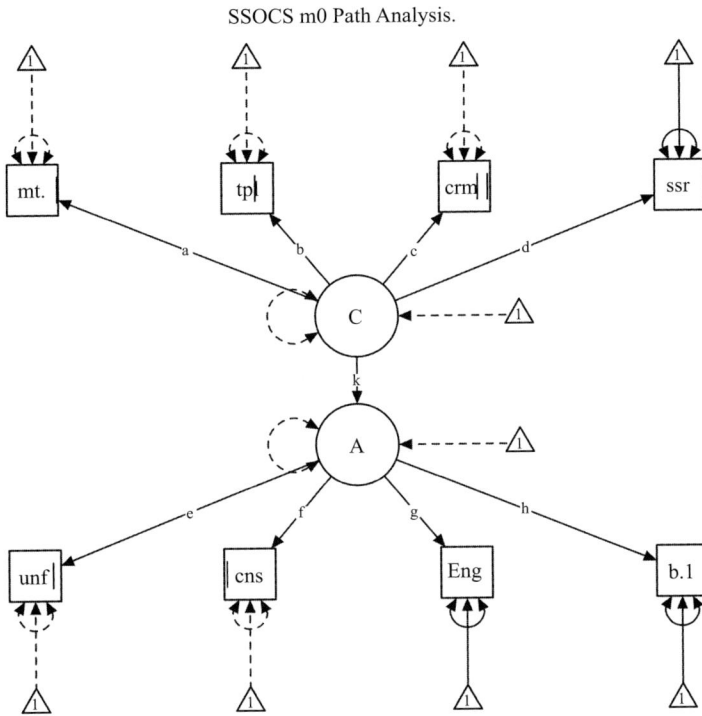

Figure 9.2 The SEM model of the SSOCS indicator variables and their respective influencing latent variables that define two latent variable models. The latent variable academic achievement (A) is assumed to covary functionally with school climate (C).

percentile on tests. The number of suspensions is severely nonnormal while the others are much less so. Natural-logarithm transformations on the suspensions variable was an ineffective remedial measure to obtain a Gaussian distribution. A zero-inflated Poisson inverse Gaussian (ZIPIG) mean model was fitted to the number of suspensions, and the residuals from this model follow a nearly Gaussian probability distribution. We therefore use these residuals in place of the suspensions in the SEM. In addition, as the latent variables are assumed to be standardized we therefore standardize the indicator variables.

Table 9.11 (reproduced from Chapter 1) gives the numbers and percentages of the categorical variables of the use of uniforms, use of metal detectors, use of tiplines, and

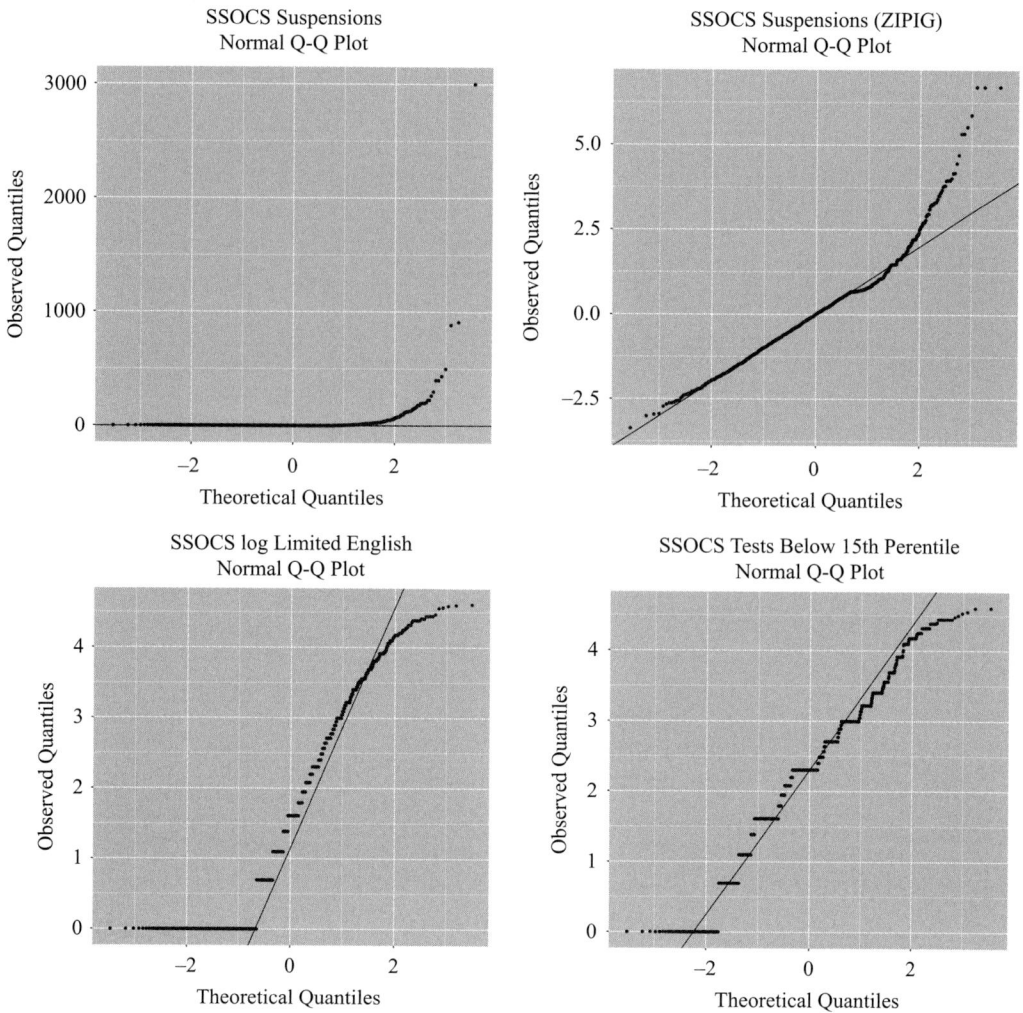

Figure 9.3 SSOCS normal Q-Q plots for the suspensions (top left-hand panel), adjusted (zero-inflated Poisson inverse Gaussian mean model residuals) suspensions (top right-hand panel), log-transformed limited English percentage (bottom left-hand panel), and the percent of students below the 15th percentile on tests (bottom right-hand panel).

use of counseling. These categorical indicator variables must be declared as categorical in the SEM analysis.

The number of parameters to estimate in the SEM is 21 as shown in Table 9.12. For large sample sizes such as we have with these SSOCS data, the sample size is multiplied by the value "fmin" to obtain a χ^2 statistic with degrees of freedom equal to the amount of nonredundant information less the number of estimated parameters. The estimated χ^2 value is 99.286 on 19 degrees of freedom and a p-value that is less than any reasonable level. This χ^2 value suggests the data are not a plausible fit to the model.

The CFI statistic has a value of 0.941. Values close to unity suggest the data are a plausible fit to the model. The RMSEA = 0.041, close to zero, suggests a good fit, and also indicates

Table 9.11 *School Survey on Crime and Safety descriptive statistics for categorical variables.*

Variable	Levels	Number	Percent
Bullying	Never	40	1.6%
	On occasion	1187	46.4%
	Monthly	547	21.4%
	Weekly	498	19.5%
	Daily	288	11.3%
Uniforms	Yes	377	14.7%
	No	2183	85.3%
Metal detectors	Yes	65	2.5%
	No	2495	97.5%
Tipline	Yes	901	35.2%
	No	1659	64.8%
Counseling	Yes	2406	94.0%
	No	154	6.0%
Discipline training	Yes	1792	70.0%
	No	768	30.0%
Behavioral training	Yes	1969	76.9%
	No	591	23.1%
Crime	Low	1922	75.1%
	Moderate	494	19.3%
	High	144	5.6%

Table 9.12 *SSOCS Model measures of fit for SEM of School Climate and Academic Achievement.*

	SSOCS SEM fit measures		
Statistic	Estimate	Degrees of freedom	*p*-value
npar	21		
fmin	0.0161		
chisq, scaled	99.2855	19	< 0.001
chisq, scaling factor	0.8342		
cfi, scaled	0.9405		
rmsea, scaled	0.0406		0.9726

the model is a reasonable approximation to the data. The RMSEA *p*-value $0.973 \gg 0.05$ is interpreted as the value of the RMSEA is at most ≤ 0.05, i.e., close to zero, which further suggests the model is a close fit to the data.

The model fit measures assess the model fit as an adequate representation of the data except the χ^2 statistic. To resolve this apparent discrepancy, we compare the correlations of the raw SSOCS in Table 9.13 with the data-model-implied correlations in Table 9.14. Table 9.13 is the lower triangular matrix of raw SSOCS data correlations. Table 9.13 is the lower triangular matrix of model-implied correlations. If the SEM is adequate,

Table 9.13　*SSOCS raw data correlation matrix for SEM of school climate and academic achievement.*

	Metal detect	Uniforms	Counsel	Tipline	Limited English	Below 15th	Crime
SSOCS raw data correlation matrix							
Metal detect							
Uniforms	0.2202						
Counsel	0.0095	0.0078					
Tipline	0.0578	−0.0085	0.0626				
English	0.0422	0.2044	0.0454	−0.0108			
15th	0.1889	0.1915	−0.0182	0.0047	0.1959		
Crime	0.1967	0.3034	0.029	0.0183	0.3176	0.3145	
Suspend	0.0285	0.0416	0.0289	0.0737	0.03	0.1122	0.1177

Table 9.14　*SSOCS model-implied correlation matrix for SEM of school climate and academic achievement.*

	Metal detect	Tipline	Crime	Suspend	Uniforms	Counsel	Limited English
SSOCS model-implied correlation matrix							
Metal detect							
Tipline	0.0325						
Crime	0.5129	0.0405					
Suspend	0.1075	0.0085	0.1341				
Uniforms	0.3971	0.0314	0.4958	0.1039			
Counsel	0.0499	0.0039	0.0623	0.013	0.049		
English	0.253	0.02	0.3158	0.0662	0.2487	0.0312	
15th	0.2723	0.0215	0.3399	0.0712	0.2677	0.0336	0.1705

the model-implied endogenous variables correlations will have approximately the same correlation values for the endogenous variables as do the raw correlations.

The model-implied correlations give the same general correlation values as the raw data correlation, except, perhaps, for those between metal detectors and limited English, metal detectors and crime, and metal detectors and uniforms. Recall that pairwise correlations are a measure of the strength of a linear relationship between two variables, so when we assess the equality of the correlations between the two correlation matrices, we are judging the relative equivalence of linearity allowing for the large size of the SSOCS sample. Considering that we posed a hypothesis for the SSOCS data that is strictly for purposes of demonstrating SEM, we would then find that the interesting difference in the correlation values is metal detectors and limited English, and uniforms. Unlike the metal detector and crime correlations, the former are under the influence of the academic achievement latent variable, and the latter is assumed influenced by the school climate latent variable. Below we will examine the regression relationship between the two latent variables, and see if this interesting correlation plays a role.

Table 9.15 *SSOCS model residuals correlation matrix for SEM of school climate and academic achievement.*

	Metal detect	Tipline	Crime	Suspend	Uniforms	Counsel	Limited English
	SSOCS model residuals correlation matrix						
Metal detect							
Tipline	0.155						
Crime	−0.014	−0.003					
Suspend	−0.043	0.085	0.014				
Uniforms	0.174	−0.048	−0.002	−0.042			
Counsel	0.005	0.166	0.015	0.046	−0.025		
English	−0.165	−0.034	0.012	−0.036	−0.007	0.084	
15th	−0.000	−0.015	−0.014	0.041	−0.035	−0.067	0.025

The residuals correlation matrix presents the among variables correlations (without the self-correlations which give no information). We look at the model residual matrix in Table 9.15 for the desirable property that no pairwise endogenous residuals are correlated. All the pairwise correlation values are less than the absolute value of 0.18 suggesting that indeed no two variable residuals pairings are correlated. Hence, we are encouraged that the model accounts for all but the random variation and the unique variation retained by the model variables.

Table 9.16 has the parameter estimates of the SEM CFA for the influence of school climate on its respective indicator variables, for the influence of academic achievement on its indicators, and the regression of the academic achievement latent variable on the school climate latent variable, all after modeling with the standardization method. The standardization method is used due to the presence of the dichotomous indicator variables. The standardization method assumes the latent variable is standardized, and thus the SEM analysis standardizes the indicator variables except for the dichotomous indicator variables as represented in the "estimate" column of the table. We see from the *p*-value column in the first nine rows of the table that all the school climate indicator variables except the tipline variable are significant at any reasonable a priori level. The school climate latent variable tends to influence the levels of its concomitant indicator variables. None of the academic achievement indicator variables have significant *p*-values, and hence, this latent variable model is likely to be mis-specified. Similarly, the regression relationship between the two latent variables is not significant which also suggests the academic achievement latent variable model is wanting. We thereby question the usefulness of this SEM example hypothesis. While the significance levels call into question the usefulness of this SEM, the loadings and communality statistics may suggest otherwise.

We calculate the communality for each indicator variable which is the portion of the indicator variance attributed to its host latent variable. Recall that communality is the squared value of the standardized parameter estimates of the indicator variable which is common to the other indicator variables with the LVM of each latent variable. The academic achievement LVM includes the communality of the school climate latent variable in addition

Table 9.16 *SSOCS parameter estimates for SEM of school climate and academic achievement path analysis model.*

		SSOCS SEM parameter estimates			
Path start	End	Estimate	Std. error	*z*-value	*p*-value
School climate	Metal detectors	0.641	0.047	13.618	0.000
	Tipline	0.051	0.036	1.423	0.155
	Crime	0.800	0.044	18.384	0.000
	Suspensions	0.182	0.029	6.248	0.000
Academic achievement	Uniforms	0.081	0.261	0.310	0.756
	Counseling	0.010	0.034	0.302	0.762
	English	0.761	2.449	0.311	0.756
	Below 15th	0.802	2.589	0.310	0.757
	Climate	7.649	25.009	0.306	0.760

Table 9.17 *SSOCS Loadings, communality, and uniqueness for SEM of School Climate and Academic Achievement.*

	SSOCS SEM variance allocation		
Variable	Loading	Communality	Uniqueness
Metal detectors	0.641	0.411	0.589
Tipline	0.051	0.003	0.997
Crime	0.800	0.640	0.360
Suspensions	0.168	0.028	0.972
Uniforms	0.625	0.390	0.610
Counseling	0.078	0.006	0.994
Limited English	0.398	0.158	0.842
Below 15th	0.428	0.184	0.816
School climate	0.992	0.983	0.017

to its indicator variables. The difference between 1 and the communality value gives the uniqueness, which is the variability not shared among the other indicator variable.

Table 9.17 gives the loadings, the communalities, and uniqueness estimates for each indicator variable and the school climate latent variable. The values of the communalities vary from essentially zero to greater than 0.4 with metal detectors at 0.411 and "crime" = 0.640. This suggests overall that the hypothesis related to school climate is at best marginally plausible. As we have seen with the other model assessments, the hypothesis associated with academic achievement is likely not plausible with uniforms = 0.390 being the largest communality. The large uniqueness values suggest the major portion of the indicator variability lies with indicator variables, and not shared relative to the associated latent variables. The communality of school climate (0.983) indicates that very little unique variance (0.017) resides with school climate, but is shared among all the indicator variables. Certainly, further consideration should be given to the hypothesis formulation if the SSOCS data are to be used to assess the influence of school climate and academic achievement on selected indicator variables. One possibility that would be interesting to pursue is using counseling as an indicator variable for both school climate and academic achievement.

Overall the fit statistics, residuals analyses, significance of the covariance parameters, and the communality values indicate this dual LVM with latent-to-latent variable regression lacks support from the SSOCS data, and hence the hypotheses are in serious question.

9.4 Summary

Structural Equation modeling (SEM) was shown to be an involved set of analytical methods with great flexibility depending on how well posed the study hypothesis is. We demonstrated the complexity and power of SEM with just two examples from the Framingham Heart Study data and the School Survey of Crime and Safety data. The Framingham hypothesis involved just one latent variable and four indicator variables with their associated error structures. The School Survey of Crime and Safety example extended the SEM from one latent variable to two latent variables each with four indicator variables, and each variable's error structure. The added regression relationship between the two latent variables took us from just latent variable modeling into the realm of full structural equation modeling, thus increasing the analytical complexity, but also adding to the model power by allowing for intricate and often convoluted dependent relationships.

The analytical power of SEM is what makes it an inviting methodological suite of analytical methods. We presented SEM in hopes that this introduction will intrigue the uninitiated analyst to explore more thoroughly the depths of hypothetical associations for informative disclosures.

9.5 Further Reading

A thorough and comprehensive text on structural equation modeling is Kline (2010). Our introduction touched upon two rather straightforward SEM models: a latent variable model and a regression model on two latent variable models, which are discussed by Kline along with several other SEM versions.

Bollen (1989) provides many SEM descriptions and examples cited in many papers. This text also is a comprehensive treatise on SEM.

Dwyer (1983) and Tabachnick and Fidell (2013) provide descriptions of SEM examples that are quite thorough in how to apply SEM.

Beaujean (2014) describes primarily latent variable modeling using R, the CRAN statistics package. The book has many worked-out examples.

10

Matching Data to Models

10.1 The Decision Process of Modeling

Given a data set or a design for collecting data, it is the task of the data analyst to match the data to an appropriate model. The selection of an appropriate model depends on a number of factors, including the goals or intentions of the study, properties of the data collected, and the nature of the conclusions the analyst would like to make. In most cases many models can be deemed appropriate for one data set, and the analyst must select one or many appropriate models to address the goals of the study. The data analyst cannot focus on "right" or "wrong" models, but must instead balance the relative strengths and weaknesses of different modeling choices. The analyst must also consider the availability of computing resources, interpretability of results, and the ability of the analyst herself.

Very generally, the data analyst must consider the specific goals or questions that need to be addressed by the study, including whether there is an interest in evaluating model effects, in making predictions using the model, or both. The analyst must also consider the nature of predictors for the analysis, including whether the predictors are continuous or categorical, whether interactions between predictors should be considered, whether any predictors should be considered as a source of random variation, whether any predictors present as time-dependent, and so on. Perhaps most relevant to the discussions from this handbook, the data analyst must consider the nature of the data collected, including whether the outcome of interest is continuous, skewed, categorical, longitudinal, or otherwise. Figure 10.1 shows some very general properties of the response of interest that must be determined by the data analyst when matching data to an appropriate model. First, the analyst must determine the type of data representing the outcome of interest which is represented by the top node, "outcome variable." Exploratory data analyses corroborate the choice of the three options for the outcome variable given in the second tier of nodes.

If the response of interest is continuous, the next step is to determine whether that continuous response is Gaussian. For Gaussian, continuous responses, the normal linear regression model is most appropriate. However, continuous responses often show non-Gaussian characteristics in the form of skewness or truncation. For skewed outcomes, it is common to apply transformations or to attempt a weighted estimation such as weighted least squares. However, we recommend moving away from such tenuous procedures and toward applying a generalized linear model appropriate for the type of data observed. For example, if an analyst is working with a response that is truncated at zero and is highly right-skewed, a gamma generalized linear model may provide superior fit as compared to a weighted least squares approach. Such options are presented in Figure 10.2.

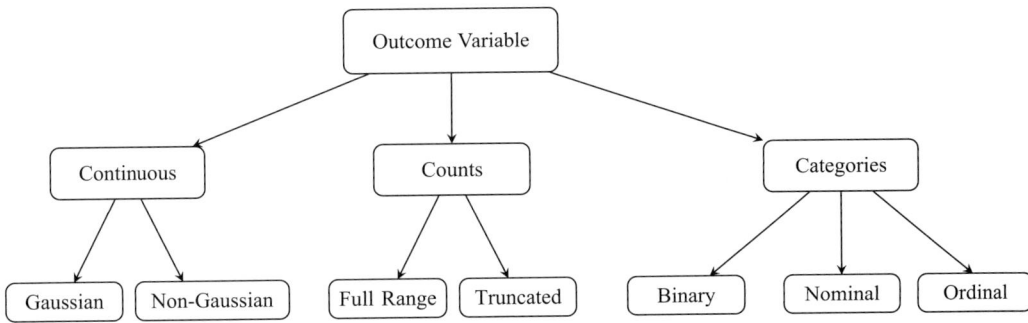

Figure 10.1 Flow chart for general decisions about response types.

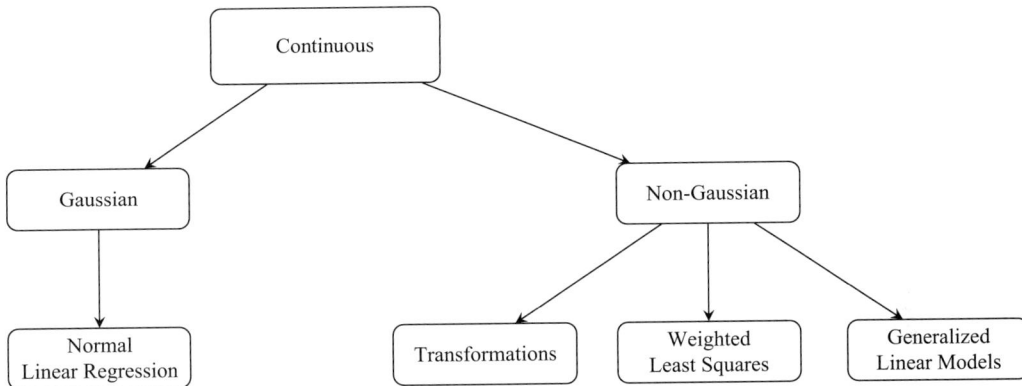

Figure 10.2 Flow chart for modeling options for continuous responses.

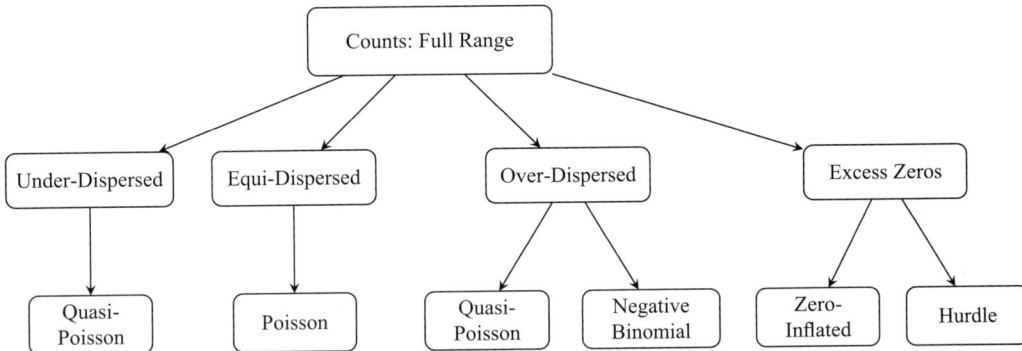

Figure 10.3 Flow chart for modeling options for count responses without range restrictions.

If the response of interest is recorded as count data, there are numerous modeling options. For ease of explanation, we have split the discussion into count data models for data that can vary across the entire range of counts (from a count of zero and without limit), and counts that exist on some restricted range. Figure 10.3 shows some decisions that an analyst must make for counts recorded on the full range of values. The most important distinction among counts is whether the variation in the count response is equal to its mean. If the mean and variation of the response are equivalent, a standard Poisson count regression model will suffice. However, this equality is rarely met in practice. When the variation exceeds the mean of the count response, the data are overdispersed and options such as the quasi-Poisson count regression or one of the many negative binomial count regression models are more appropriate than the standard Poisson model. The quasi-Poisson model can also accommodate underdispersed data, although this situation is less prevalent than overdispersion.

An increasingly common data situation is that of count responses with excess zeros, in which a greater proportion of zero counts have been observed than would be expected under the assumed distribution, usually Poisson or negative binomial. Excess-zero-modeling options fall generally into two categories: zero-inflated models and hurdle models. Zero-inflated models allow zeros from two separate processes such that with a certain probability there will be zeros from one sub-population of the study, while the other sub-population may happen to produce zero counts on occasion. Hurdle models impose the assumption that zero counts come from a process different than the process for the positive counts, and that the population of zeros is distinct from the population of the positive counts.

When counts are observed on a restricted range of possible values, the equal dispersion assumption of the Poisson count regression model cannot be met, as the data are inherently underdispersed. Figure 10.4 shows some decisions the data analyst must make regarding such restricted counts. The most common type of restriction in practice is the zero-truncated count, for which the count of zero is not possible but all other counts are possible. It is also possible to have counts left-truncated by any lower bound, right-truncated by any upper bound, or double-truncated by both bounds. It is important for the analyst to recognize that these restrictions are not observed through the data, but rather by the data collection methods. A count response without any zeros is not necessarily zero-truncated; it is possible that the data simply happen not to have any observations of zero counts. An outcome of interest is zero-truncated only if it is *impossible* to observe a count of zero.

If the response of interest represents mutually exclusive categories, logistic models are generally preferred. For this type of data, the analyst must determine the number of possible outcome categories, and whether these categories can be ordered in some meaningfully quantitative way. For the simplest case of two possible outcome categories, binary logistic regression is preferred to model the probability associated with one of the outcome categories. Other options have been developed, including the probit regression

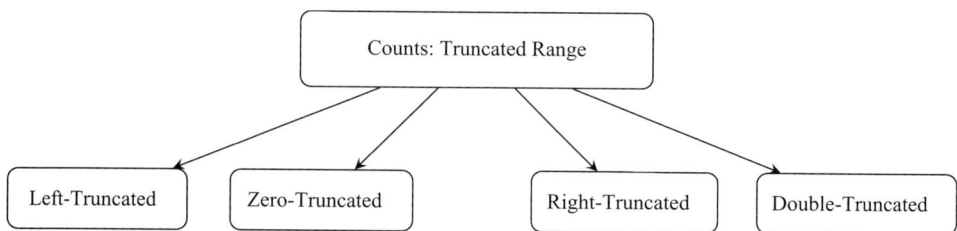

Figure 10.4 Flow diagram for modeling options for restricted counts.

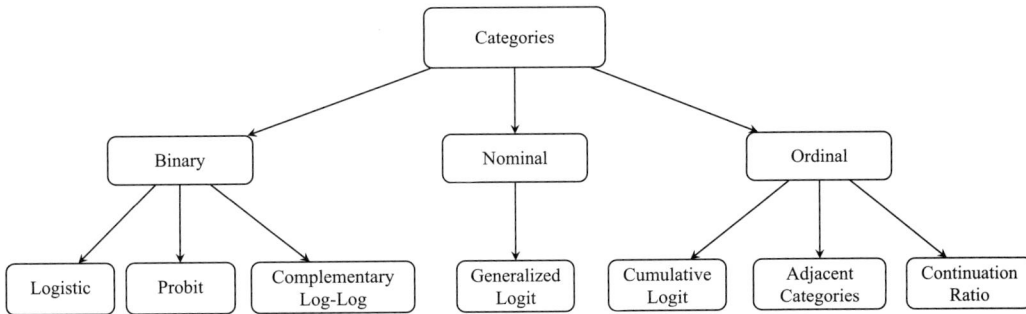

Figure 10.5 Flow chart for modeling options for categorical responses.

model and the complementary log-log regression model. However, we prefer the logistic regression model because it generally results in the most powerful effects tests and produces interpretations most easily connected to relevant statistics such as odds and probabilities.

Figure 10.5 shows modeling options for categorical outcomes, including those with more than two options. If the response categories do not include any inherent ordering, the data are said to be nominal and the basic generalized logit model is most common, and estimates odds associated with each category as compared to an arbitrarily selected baseline category. If the response categories can be ordered in a meaningful way, the data are said to be ordinal, and many modeling options have been developed. While the cumulative-logit model, in which outcome categories are collapsed to a single lower category and a single upper category, remains the most commonly available option, the other models can provide interpretations of interest in specific modeling situations. The adjacent-categories model allows for easy calculation of odds comparing consecutive outcome categories, and the continuation-ratio model allows the calculation of odds comparing each outcome category to all categories greater or lesser.

If the outcome of interest has been recorded repeatedly over a number of times of observation, all of the above models can be extended to longitudinal versions, shown in Figure 10.6. When applying a longitudinal model, a marginal approach can be used to obtain population-averaged conclusions about general associations while properly accounting for the inherent autocorrelation in the responses. Marginal methods of estimation such as the generalized estimating equations allow the data analyst to assume any type of correlation among the outcomes over time.

In order to make conclusions about individual growth or change over time, the subject-specific conclusions of a conditional longitudinal model will be more appropriate. In order to apply a conditional longitudinal model, it is necessary for the analyst to identify variables that represent sources of random fluctuation in the outcome or in associations between predictors and the outcome. Data analysts often rely on the assumption of exchangeable autocorrelation, meaning that the associations between responses do not depend on the time lag between the responses. The most straightforward way to impose this exchangeable assumption is to apply a random-intercept model, in which the baseline value fluctuates randomly among the population of interest. Other types of autocorrelation can be imposed using more complicated models which include interactions between predictors and

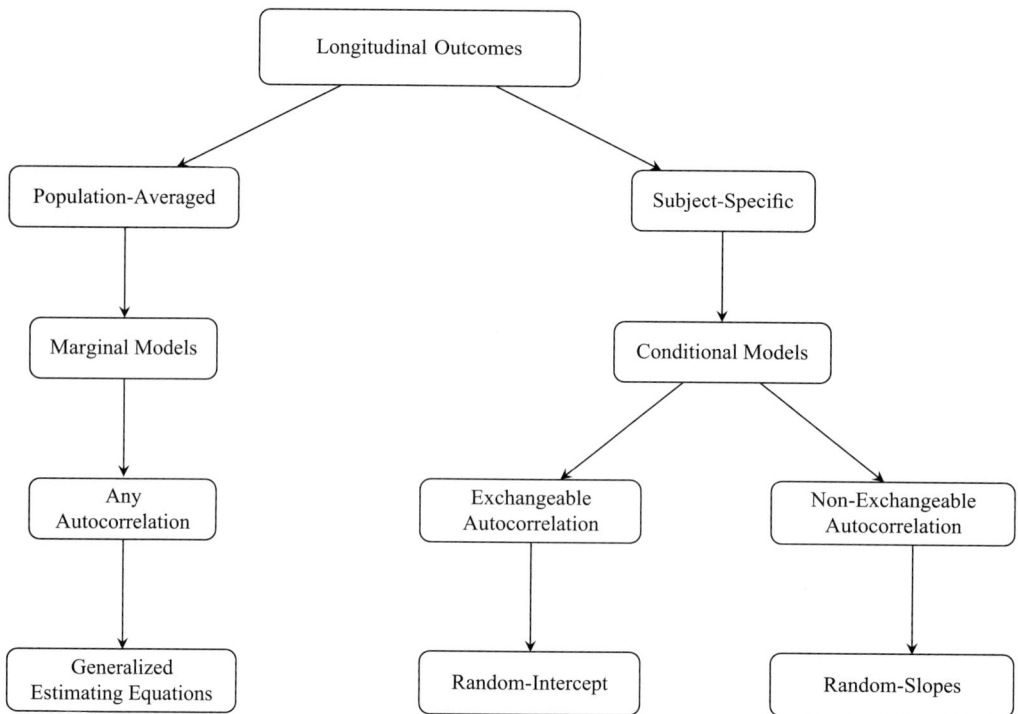

Figure 10.6 Flow chart for modeling options for longitudinal responses.

sources of random variation, called random-slopes models. In such models the effects can fluctuate randomly within the population of interest.

When the response of interest represents the time to an event, a separate class of models referred to as time-to-event, survival, or failure models is appropriate. For such models the data analyst must determine whether the predictors include time-dependent covariates, whether the relationships between any predictors and the rate of occurrence of the event of interest change over time, and whether the analyst is interested in the first occurrence of the event or in multiple occurrences of the event. The data analyst must also consider censoring of individuals, in which the exact time of the event is not observed during the study but nonevent information can be used.

For latent variables, that is, variables that are not observed directly, proxy variables or observed variables that are influenced by the latent variable are used in the class of models referred to as structural equation models. This class of models allows the latent variable to be influenced by observed variables, or the latent variable may be thought to drive a set of observed variables. Additionally, a regression model with latent variables as responses may be influenced by one or more driving latent variables. In each situation, the latent variable behavior is assessed by concomitantly observed variables.

There are a number of modeling considerations that we were not able to cover in this handbook. This discussion should be treated as a starting place for proper data analysis, but not as a complete reference. In particular, we omitted any discussion of the calculation of effect sizes, estimation techniques that account for predictor multicollinearity, the use of sampling weights, or adjustments for nonconvergence of standard estimation techniques. We did not discuss the full spectrum of possible generalized linear models, recurrent events analysis, time series, or Bayesian techniques for estimation and hypothesis testing. However, all of the above-mentioned topics can be investigated further using the listed references.

10.2 Results of Model Application

Throughout this handbook we fit a number of models using the same four data sets. None of the fitted models should be considered as right or wrong models, but rather as different options with relative strengths and weaknesses. The outcomes of our analyses are for demonstrative purposes only, and should not be construed as validated research findings.

10.2.1 School Survey on Crime and Safety

We used the 2007–2008 School Survey on Crime and Safety to model the number of suspensions for insubordination, the prevalence and likelihood of different types of bullying, and general student behavioral problems.

We modeled the number of suspensions using a weighted least squares adjustment to a normal linear model. This approach allowed us to continue to use familiar linear regression methods, while by supposition accounting for the changing variation of the number of suspensions. However, this method was not particularly effective in accounting for the nonconstant variation and is not preferred. We saw that the selection of weights was arbitrary, and results depended on which weights we chose to apply. In a more appropriate analysis of the number of suspensions, we applied a hurdle negative binomial count regression model. This analysis in particular allowed us to determine which factors were associated with the likelihood of a school reporting zero suspensions, and also which predictors had a significant association with higher numbers of suspensions. This model accounted for the particular distributions of suspensions, which included overdispersion and a majority of zero observations.

We modeled the likelihood of reporting different levels of bullying using an adjacent categories multinomial logistic regression model. This model was selected because the outcome of interest was recorded as multiple classifications with an inherent ordering. The model allowed us to make conclusions about the probability associated with each level of bullying for different types of schools.

A structural equation model example was constructed to demonstrate how to model two latent variables each with four indicator variables, and with the added regression relationship between the two latent variables to form a full structural equation model. The hypothesis tested was that academic achievement, as indicated by the use of uniforms, use of counseling, English language competency, and the prevalence of test results below the 15th percentile, is influenced by the school climate as indicated by the use of metal detectors, tip lines, prevalence of crime, and the number suspensions. We saw that academic achievement was not significantly influenced by school climate under these model conditions. However, we were able to see how to construct and analyze a structural equation model, and the flexibility these models can bring to complex processes.

10.2.2 Framingham Heart Study

The Framingham Heart Study was used to model the chances of showing evidence of hypertension, time to hypertension, and also to model general stress for individuals.

We modeled the likelihood of hypertension using logistic regression models because presence of hypertension is a simple yes/no binary response. We modeled the probability of showing hypertension at any time in the Framingham study as a Cox regression survival model, and we also applied a marginal longitudinal model to predict the likelihood of showing hypertension at different periods of observation. The marginal longitudinal model will generally be more powerful than the simple logistic regression model because more outcomes are considered, as responses are taken at each observation time. In both cases we were able to discuss the likelihood and the odds of showing hypertension for different populations, and also to report odds ratios comparing the odds across populations. The Cox regression model allowed us to consider censored data as well as report the influence of predictors on the time to hypertension.

The latent variable stress as indicated by four observed variables is an example of a latent variable model, and is a straightforward application of a structural equation model. The Framingham data were used to assess the hypothesis that the latent variable, stress, is consequential to the four indicator variables. The fit statistics suggested the stress latent variable model was an influencer only on the onset of hypertension. If this model were of serious consideration, the conclusion is that the stress model would need reconsideration.

10.2.3 Fire-Climate Interactions in the American West

We used the Fire-Climate Interactions data to model decade fire counts and counts averages between the decades of 1130 and 2000, and compared these values across the Pacific Northwest, Northern California, Southwest, and Interior West regions of the American West.

We modeled decade fire averages using a weighted least squares approach, using decade variances as weights. For such aggregated data it is common to apply a weighted approach to account for the nonconstant variation inherent in the data. However, this approach did not adequately account for all of the characteristics of the data and is not recommended.

We applied a zero-inflated negative binomial counts regression model on the decade counts, which was far superior to the weighted least squares approach. After trying Poisson and negative binomial regression models, we chose the zero-inflated negative binomial model that allowed us to make conclusions about which regions showed different chances of showing zero counts of fire indicators, and also which regions had greater expectations for the number of decade fire counts.

In order to predict trends of fire counts across decades, we applied a conditional longitudinal Poisson count regression model, allowing the baseline decade counts of fire indicators to fluctuate randomly among sites within three regions. As part of this model, we included a quadratic relationship between fire count and time, allowing the model to capture the changing trends in fire counts. This conditional approach allowed us to construct a prediction function for each region across the decades of observation, while accounting for the inherent autocorrelation in fire counts among sites within the same region. While we implicitly applied an exchangeable autocorrelation structure through the random-intercept model, other autocorrelation structures were investigated through random-slopes models.

10.2.4 English Wikipedia Clickstream

The English Wikipedia Clickstream data were used to investigate the behavior of pairs of referring and requested pages during an approximately three-hour span from the beginning of February, 2015. We considered the factor of the previous site requesting the Wikipedia page.

We modeled the probability of requesting a redlink, or in other words a request that could not be completed. We applied a logistic regression model because of the binary yes/no nature of the response. We saw that some of our traditional fit statistics behaved differently for such a large data set and consequently we had to consider other methods to assess model quality. Model predictions showed extremely small likelihoods of a redlink, which is common for a logistic model of a rare event.

The clickstream data are left-truncated with counts of 10 and larger. We adjusted the counts probability distribution to exclude the nonexistent data (0 through 9) to utilize a truncated count model. After testing several truncated distributions, the Sichel 2 distribution, which is particularly useful for highly overdispersed count data such as is the case with clickstream data, was found to give the best-fitting model. The model fit statistics and diagnostic plots revealed that the truncated count model was not adequate. While the model fit overall was inadequate, the Sichel 2 truncated count model gives only 0.58% of predicted counts under 10. There are only 0.09% predicted counts fewer than 8. This is a reasonable outcome considering the model construction data have no counts below 10 relative to the far-inferior normal linear model which gave a fitted range from 21 to 56, a severe distortion of the observed counts.

10.3 Perspectives on Modeling

The data analyst has a stimulating and challenging task in matching a given data set to an appropriate model or combination of models. We believe there is an almost overwhelming amount of literature available to describe various modeling options. Ultimately the analyst must select a data modeling approach that will lead to the most informative conclusions with respect to the intention of the analysis.

We have made a few general observations about the development of statistical modeling that represent some best practices for data analysis. We believe analysts should strive to use the simplest model possible that is appropriate and answers the questions of interest. This is not an endorsement of the normal linear regression model, but rather a caution against pursuing a model that needlessly accounts for every nuance of the data under consideration. This choice has a subjective element, and often the analyst must make difficult decisions in balancing complexity with appropriateness. But ultimately all models involve concessions in which some properties of the data must be ignored, so there is a place for allowing known imperfections in selected models. Best practice dictates these imperfections be clearly identified and documented.

We believe the use of data transformations to accommodate the normal linear model is an outdated approach that should no longer be applied. In short, this approach creates more problems than it solves, as interpretations of model coefficients become challenging, and the correction of one failed assumption frequently leads to the failure of another assumption. We believe data analysts should instead turn to models that transform the *parameters* of interest, not the data. For example, within count regression models the mean or rate is transformed, usually through a natural-logarithm transformation. The data remain unchanged, but the mean parameter is transformed to interact with predictors and parameters on something similar to a linear scale. In this way interpretations remain consistent, and a model is applied that is appropriate for the specific data type under consideration. Given current software options, the use of, for example, logistic and count regression models is just as straightforward as normal linear regression. There is no reason to turn away from an easily available tool that is more appropriately constructed for the outcome data type, provided the analyst clears the hurdle of understanding the meaning of effects coefficients.

Most statistical analysis packages have what are intended to be user-friendly interfaces for guiding the construction of models. These interfaces, point-and-click, for example, often prevent the user from applying techniques not suited to the data and model combination of choice. Some interfaces make recommendations for model construction and effects testing. Point-and-click operations may seem an advance, but the analyst must not rely on software to produce adequate models for a particular data set as the software cannot know the intent of the study. Only the analyst can synthesize the study purpose with the data to produce a model from which insightful knowledge results.

The modeling techniques discussed in this handbook should be treated as an introduction and the applications of these models should be used as guidelines on the use and interpretation of such models. This handbook does not make the reader an expert in the material, but provides an accessible starting place from which the references can take the analyst further. In many modeling situations it can be beneficial to contact a statistician to ensure methods have been applied properly and all appropriate methods have been explored. Data modeling provides accessibility to information often confounded by variable types and data organization within a given data set. While care and subject matter expertise are critical to constructing consummate models, once a model befits the data and the study intention, the beneficiary is the body of knowledge.

Bibliography

Agresti, A. (1998). *Categorical Data Analysis*, 3rd edn, John Wiley & Sons, Inc.

Agresti, A. (2007). *An Introduction to Categorical Data Analysis*, 2nd edn, Wiley Interscience.

Agresti, A. (2010). *Analysis of Ordinal Categorical Data*, 2nd edn, Wiley.

Agresti, A. and Finlay, B. (2008). *Statistical Methods for the Social Sciences*, 4th edn, Pearson.

Allison, P. D. (2010). *Survival Analysis Using SAS: A Practical Guide*, 2nd edn, SAS Institute.

Allison, P. D. (2012). *Logistic Regression Using SAS: Theory and Application*, SAS Institute Inc.

Beaujean, A. A. (2014). *Latent Variable Modeling Using R: A Step-by-Step Guide*, Taylor & Francis.

Bentler, P. M. and Bonett, D. G. (1980). Significance tests and goodness of fit in the analysis of covariance structures. *Psychological Bulletin*, 88, 588–606.

Bollen, K. A. (1989). *Structural Equations with Latent Variables*, John Wiley & Sons, Inc.

Cameron, A. C. and Trivedi, P. K. (2013). *Regression Analysis of Count Data Book*, 2nd edn, Cambridge University Press.

Cleveland, W. S. (1993). *Visualizing Data*, Hobart Press.

Collett, D. (2003). *Modelling Binary Data*, Chapman & Hall.

Cox, D. R. and Snell, E. G. (1989). *Analysis of Binary Data*, 2nd edn, Chapman and Hall.

Crawley, M. J. (2007). *The R Book*, 1st edn, Wiley Publishing.

Diggle, P. J., Heagerty, P., Liang, K. Y., and Zeger, S. L. (2002). *Analysis of Longitudinal Data*, 2nd edn, Oxford University Press.

Dobson, A. J. (2002). *An Introduction to Generalized Linear Models*, 3rd edn, Chapman & Hall/CRC.

Dwyer, J. H. (1983). *Statistical Models for the Social and Behavioral Sciences*, Oxford University Press.

Ebeling, C. E. (2004). *An Introduction to Reliability and Maintainability Engineering*, McGraw-Hill.

Faraway, J. J. (2006). *Extending the Linear Model with R: Generalized Linear, Mixed Effects and Nonparametric Regression Models*. CRC Press.

Fitzmaurice, G. M., Laird, N. M. and Ware, J. H. (2012). *Applied Longitudinal Analysis*, 2nd edn, John Wiley & Sons.

Fitzmaurice, G., Davidian, M., Verbeke, G. and Molenberghs, G. (2008). *Longitudinal Data Analysis*, 1st edn, Chapman & Hall/CRC.

Freedman, D. A. (2009). *Statistical Models: Theory and Practice*, 2nd edn, Cambridge University Press.

Freedman, D. A., Pisani, R., and Purves R. (2007). *Statistics*, 4th edn, W. W. Norton & Company.

Glass, G. V. and Hopkins, K. D. (1995). *Statistical Methods in Education and Psychology*, 3rd edn, Allyn & Bacon.

Hardin, J. W. and Hilbe, J. M. (2003). *Generalized Estimating Equations*, Chapman & Hall.

Hedeker, D. and Gibbons, R. D. (2006). *Longitudinal Data Analysis*, Wiley Interscience.

Heinze, G. and Schemper, M. (2002). A solution to the problem of separation in logistic regression. *Statistics in Medicine*, 21, 2409–2419, 2002.

Hilbe, J. M. (2011). *Negative Binomial Regression*, Cambridge University Press.

Hilbe, J. M. (2014). *Modeling Count Data*, Cambridge University Press.

Hilbe, J. M. (2015). *Practical Guide to Logistic Regression*, Taylor & Francis.

Hosmer, D. W., Lemeshow, S., and May, S. (2008). *Applied Survival Analysis*, 2nd edn, Wiley Interscience.

Hosmer, D. W., Lemeshow, S., and Sturdivant, R. X. (2013). *Applied Logistic Regression*, 3rd edn, Wiley.

Kleinbaum, D. G. and Klein, M. (2010). *Logistic Regression: A Self-Learning Text*, 3rd edn, Springer.

Kline, R. B. (2010). *Principles and Practice of Structural Equation Modeling*, 3rd edn, Guilford Press.

Kutner, M., Nachtsheim, C., Neter, J., and Li, W. (2004). *Applied Linear Statistical Models*, 5th edn, McGraw-Hill/Irwin.

Lee, Y., Nelder, J. A. and Pawitan, Y. (2006). *Generalized Linear Models With Random Effects*, Chapman & Hall/CRC.

McCullagh, P. and Nelder, J. A. (1989). *Generalized Linear Models*, 2nd edn, Chapman & Hall/CRC.

McKillup, S. (2011). *Statistics Explained: An Introductory Guide for Life Scientists*, 2nd edn, Cambridge University Press.

Meeker, W. Q. and Escobar, L. A. (1998). *Statistical Methods for Reliability Data*, Wiley.

R Core Team (2016). *R: A Language and Environment for Statistical Computing*, R Foundation for Statistical Computing. www.R-project.org.

Snedecor, G. W. and Cochran, W. G. (1989). *Statistical Methods*, 8th edn, Iowa State University Press.

Tabachnick, B. G. and Fidell, L. S. (2013). *Using Multivariate Statistics*, 6th edn, Pearson Education.

Tobias, P. A. and Trindade, D. (1995). *Applied Reliability*, 3rd edn, Taylor & Francis.

Trouet, V., Taylor, A. H., Wahl, E. R., Skinner, C. N., and Stephens, S. L (2010). Fire-Climate Interactions in the American West Since 1400 CE. *Geophysical Research Letters*, 37(4). DOI: http://10.1029/2009GL041695.

Tukey, J. W. (1977). *Exploratory Data Analysis*, Addison-Wesley Publishing Company.

Twisk, J. W. R. (2006). *Applied Multilevel Analysis: A Practical Guide for Medical Researchers*, 1st edn, Cambridge University Press.

Twisk, J. W. R. (2013). *Applied Longitudinal Data Analysis for Epidemiology: A Practical Guide*, 2nd edn, Cambridge University Press.

Verzani, J. (2014). *Using R for Introductory Statistics*, 2nd edn, Chapman & Hall/CRC.

Vuong, Q. H (1989). Likelihood ratio tests for model selection and non-nested hypotheses. *Econometrica*, 57(2), 307–333. www.jstor.org/stable/1912557.

Wood, S. (2006). *Generalized Additive Models: An Introduction with R*, CRC Press.

Wulczyn, E. and Taraborelli, D. (2015). *Wikipedia clickstream*, February, 2015. https://figshare.com/authors/Ellery_Wulczyn/689409.

Zeger S. L. and Liang, K. Y. (1992). An overview of methods for the analysis of longitudinal data. *Statistics in Medicine*, 11(14–15), 1825–1839.

Ziegler, A. (2011). *Generalized Estimating Equations*, 1st edn, Springer.

Index

adjacent categories ordinal logistic models, 83
Akaike information criterion (AIC), 44, 45,
    117–120, 146
autocorrelation, 152, 153, 157
    autocorrelation function, 154
    autocorrelation plots, 155
    stationarity, 155
    variogram, 155
    working correlation structure, 157

bar plots, 78
Bayesian information criterion (BIC), 45
binary outcomes, 76
box plots, 58
Breusch-Pagan test, 59

categorical models, 76, 82–84, 205
categorical outcomes, 76, 205
censored data, 38, 132–134, 150
coefficient of determination (R^2), 43, 44
conditional longitudinal models, 39, 156, 160, 208
confirmatory factor analysis (CFA), 184, 187, 192,
    199
constant variance models, 50
contingency tables, 77
continuation ratio ordinal logistic models, 84
count models, 35, 44, 109, 111, 112, 130, 203
    excess-zeros, 36, 122
    hurdle, 37, 115, 122–124, 130
    left-truncated, 209
    negative binomial, 36, 111, 113–115, 118, 119,
        122, 123, 130
    Poisson, 35, 109, 110, 113, 114–119, 122, 130
    Poisson regression, 109
    quasi-Poisson, 36, 110, 113, 117, 118, 122
    truncated, 123, 126, 128, 129
    zero-deflated, 37, 114
    zero-inflated, 36, 114, 115, 119, 120, 121, 122,
        130
Cox models, 45, 148
    proportional hazard (CPH), 134, 135, 140, 142,
        143, 146–148, 150
    proportional hazards regression, 38, 39

time-dependent, 145–149
cumulative logit ordinal logistic models, 83

data sets
    English Wikipedia Clickstream data, 3, 20, 34,
        37 55, 101, 126, 128–131, 209
    Fire-Climate data, 32, 53–55, 70, 115, 116, 118,
        119, 122, 130, 131, 172, 184, 208
    Fire-Climate Interactions, 2, 17, 37, 41
    Framingham Heart Study (FHS), 2, 13, 34, 39,
        41, 42 52, 53, 85, 135, 138, 142, 146, 150,
        163, 186, 189–191, 193, 194, 201, 208
    School Survey on Crime and Safety (SSOCS), 2,
        8, 32, 34, 37, 42, 50, 61, 93, 124, 130, 131,
        194–201, 207
deviance, 44, 45, 79, 117, 118, 162
dispersion parameter, 113

effect size, 43
effects analysis, 29
endogenous variables, 184–186, 192, 198, 199
English Wikipedia Clickstream data, 3, 20, 34, 37,
    55, 101, 126, 128–131, 209
estimation
    iterated weighed least squares, 79
    maximum likelihood, 79
excess-zero, 204
excess-zero models, 36, 114, 122
exogenous variables, 184–186
exploratory data analysis (EDA), 3, 26, 56, 115,
    122, 123, 126, 134, 185
    bar plots, 78
    contingency tables, 77
    interaction plots, 154
    logistic histogram plot, 77
    spaghetti plots, 153
    time plots, 153
    variogram, 155
    box-whisker plots, 5
    data summary and tables, 4
    graphics, 4
    histograms, 5
    loess curve, 6
    pairwise correlation, 7

213

exploratory data analysis (EDA), (*cont.*)
Q-Q plots, 5
scatter plots, 6

factor analysis, 187
failure analysis, 133
failure models, 37
Fire-Climate Interactions, 2, 17, 32, 37, 41, 53–55, 70, 115, 116, 118, 119, 122, 130, 131, 172, 184, 208
Framingham Heart Study (FHS), 2, 13, 34, 39, 41, 42, 52, 53, 85, 135, 138, 142, 146, 150, 163, 184, 186, 189–191, 193, 194, 201, 208

Gauss-Markov assumptions, 4, 54, 56, 109
Gaussian models, 50, 52–56, 133, 185
generalized estimating equations (GEE), 40, 45, 157

hazard function, 133
heteroscedasticity, 51, 52, 54, 55, 57, 58
homoscedasticity, 51, 52, 54, 56, 57
Hosmer-Lemeshow test, 80
hurdle models, 37, 109, 115, 122–124, 130, 204

indicator variables, 184, 186, 190, 192–195, 199
information criteria, 44, 162
Akaike information criterion (AIC), 162
Bayesian information criterion (BIC), 162
interaction plot, 154
iterated (re-)weighted least squares (IWLS or IRLS), 27

Kaplan-Meier method (K-M), 134, 135, 137–140, 144, 145, 150

latent variable models (LVM), 184–187, 193, 194, 199, 201
latent variables, 183, 186–188, 190, 192, 194, 195, 199
Levene's test, 58
life tables, 133–138, 140, 145, 150
likelihood function, 33
logistic histogram plot, 77
logistic models, 32, 44, 76–78, 114, 132, 140, 142, 204, 208, 209
adjacent categories, 83, 205
binary logistic, 77, 204, 208, 209
continuation ratio, 84, 205
cumulative logistic, 205
cumulative logit, 83
logit, 78
nominal multinomial logistic, 82, 205
odds, 78, 82, 84
odds ratio, 79

ordinal multinomial logistic, 83, 207
longitudinal data, 39, 152
longitudinal models, 45, 153, 205
conditional longitudinal, 156, 160, 205
conditional longitudinal models, 208
generalized estimating equations (GEE), 157, 205
marginal longitudinal, 156, 205
population-averaged, 156
random effects, 160
random-intercept, 160
random-slopes, 161
residuals, 162
subject-specific, 156
transition, 156
working correlation structure, 158

machine learning
big data, 7
cluster analysis, 7
data mining, 7
lift charts, 7
nearest neighbors, 8
neural networks, 8
random forests, 7
regression trees, 8
support vector machines, 8
Mahalanobis distance, 142
manifest variables, 183, 186–188
marginal longitudinal models, 39, 156, 157
maximum likelihood, 27, 33, 35, 41, 109
model fit, 43, 45, 143, 146, 188, 189, 191, 193, 196, 197, 201
Akaike information criterion (AIC), 117–120, 146
coefficient of determination (R^2), 43, 44
comparative fit index (CFI), 188, 189, 191
deviance, 44, 45, 118
log rank score test, 45
Pearson χ^2, 44, 45, 111, 113, 118, 188, 189, 191
pseudo R^2, 44, 79
residuals, 45, 47, 192, 199, 201
root mean square error of approximation (RMSEA), 188, 189, 191
root mean square residual (RMR), 189
standardized root mean square residual (SRMR), 188, 189, 191
Vuong's test, 120, 122
Model Types
adjacent categories, 83, 205
binary logistic, 204, 208
conditional longitudinal, 156, 160, 205, 208
continuation ratio, 84, 205
counts, 203, 209
count regression, 36, 37, 34–36, 44
Cox models, 148

cumulative logistic, 205
cumulative logit, 83
excess zeros, 122, 204
failure analysis, 133
generalized estimating equations (GEE), 40, 157, 205
hurdle, 109, 115, 122–124, 130, 204
logistic, 44, 76, 204
logistic regression, 32, 78, 132, 140, 142
logit, 78
longitudinal, 39, 45, 205
marginal longitudinal, 156, 205
negative binomial, 109, 111, 113–115, 118, 119, 122, 123, 130, 204
nominal multinomial logistic, 82, 205
nominal multinomial logistic regression, 34
nonconstant variance, 31
normal linear regression, 30, , 43, 50, 52–56, 124, 130, 133, 185
ordinal multinomial logistic, 83, 207
ordinal multinomial logistic regression, 34
Poisson, 110, 113–119, 122, 130
Poisson count, 204
Poisson regression, 109, 110
quasi-Poisson, 110, 113, 117, 118, 122, 204
random effects, 160
random-intercept, 160
random-slopes, 161
seemingly unrelated regression (SUR), 42
structural equation modeling (SEM), 41, 183–185, 192, 196–198, 206–208
survival analysis, 133
survival models, 132
time-to-event (TTE), 37–39, 45, 133–140, 142–150, 206
transformations, 59, 202
transition, 156
truncated, 123, 126, 128, 129
truncated count, 204
weighted least squares, 31, 60, 125, 130, 202, 208
zero-deflated, 109, 114
zero-inflated, 109, 114, 115, 119, 120, 121, 122, 130, 204, 208
model-buiding, 26, 202
  construction, 27
  diagnostics, 28
  effects analysis, 28, 29
  estimation, 27
  interpretation, 29, 111–113, 115
  parameter estimation, 27, 111
  prediction, 29
  specification, 27

negative binomial models, 36, 109, 111, 113–115, 118, 119, 122, 123, 130
nominal multinomial logistic models, 82

nominal multinomial logistic regression, 34
normal linear regression, 30, 31, 34, 38, 43, 46, 50, 52–57, 76, 124, 130, 133, 152, 185, 202

odds, 78, 82–84
odds ratio, 79
ordinal multinomial logistic models, 83, 207
ordinal multinomial logistic regression, 34
ordinary least squares, 27, 30, 185, 108
overdispersed, 128
overdispersion, 36, 109–111, 113, 116, 118, 123

p-value, 28, 111, 113, 118
panel data, 152
Pearson χ^2, 44, 45, 113, 118, 146, 188, 191
Poisson count models, 110, 114
Poisson models, 35, 109, 110, 113, 115–119, 122, 130
Poisson regression, 109
population-averaged, 39, 156
pseudo R^2, 44, 79

quasi-likelihood under the independence model criterion (QIC), 45, 159
quasi-Poisson, 113
quasi-Poisson count regression, 36
quasi-Poisson models, 110, 113, 117, 118, 122

random effects, 113, 160
random-index model, 40
random-intercept model, 160
random-slopes models, 161
receiver operating characteristic, 81
residuals, 45, 81, 111 , 124, 135, 159, 162, 185, 187, 192, 199, 201
  conditional longitudinal model residuals, 162
  conditional residuals, 47
  deviance, 46, 79, 162
  linearized-scale residuals, 47
  marginal residuals, 47
  Pearson residuals, 46
  plots, 58, 61
  Schoenfeld residuals, 47, 144, 145

Schoenfeld residuals, 135, 144, 145
School Survey on Crime and Safety (SSOCS), 2, 8, 32, 34, 37, 42, 50, 61, 93, 123, 124, 130, 131, 184, 194–201, 207
seemingly unrelated regression (SUR), 42
sensitivity, 81
separation, 77
spaghetti plot, 153
specificity, 80
stationarity, 155
structural equation modeling (SEM), 41, 183–185, 188, 192, 196–198, 201, 206, 207

structural equation modeling (SEM), (*cont.*)
 communality, 187, 188, 190, 193, 199, 200
 confirmatory factor analysis (CFA), 192, 199
 endogenous variables, 184–186, 192, 198, 199
 exogenous variables, 184–186
 factor analysis, 187, 188
 indicator variables, 184–186, 190, 192–195, 199
 latent variable models (LVM), 184–187, 193, 194, 199, 201, 208
 latent variables, 183, 184, 186–188, 190, 192, 194, 195, 199
 manifest variables, 183, 184, 186–188
 path analysis, 184–187, 194
 regression component, 184
 seemingly unrelated regression (SUR), 185
 structural component, 184
 unique variance, 188, 191, 193, 200
subject-specific, 40, 156
survival analysis, 133
survival models, 37, 132, 150

time plot, 153
time-to-events (TTE), 37, 132, 206
 analysis, 132–134

Cox proportional hazards, 38, 39, 45
data, 132
Kaplan-Meier method (K-M), 134, 135, 137–140, 144, 145, 150
life tables, 133–138, 140, 145, 150
mean cumulative function (MCF), 38
models, 132–134
transition models, 156
truncated count models, 204
truncated models, 123, 126, 128, 129

variance-stabilizing transformation, 54, 55, 59, 109, 123, 124, 140, 144
variogram, 155
Vuong's test, 120, 122

weighted least squares, 31, 46, 60, 125, 130, 208
White's test, 59
working correlation structure, 158

zero-deflated models, 37, 114, 109
zero-inflated models, 36, 109, 114, 115, 119, 120, 121, 122, 130, 204, 208